You picked up this book because your head aches, your heart aches, and your stomach aches. Day in and day out you regret what you do and for whom you do it. Deep down, you know you have more to contribute to the world. You see the signs in your coworkers and neighbors: the glazed-over eyes and heavy feet of drudgery.

YOU'RE NOT ALONE.

For seventeen years, 70 percent of the American workforce has remained disengaged. With every tick of the clock, tens of millions of people anxiously sleepwalk another inch closer to their breaking point. Stock options and free drinks in the break room haven't moved the needle. Our employers are failing us, but it's not their fault: they aren't designed to light the fire inside us. We have a societal problem. We're more talk than action. We allow inertia to rule our day. We rely on myths instead of science. But we are not beyond hope. Reshaping our work culture is possible!

IT FALLS ON US, AND THE TIME IS NOW.

ISBN 978-1-5445-1042-2

SHIFT THE WORK

TO ELIANA AND JAMES

REMEMBER, YOU CAN *ALWAYS* FIND A WAY.

BEFORE WE DIG INTO THIS CONVERSATION, THERE'S SOMETHING ABOUT YOUR BRAIN YOU NEED TO KNOW.

YOU HAVE THREE OF THEM.

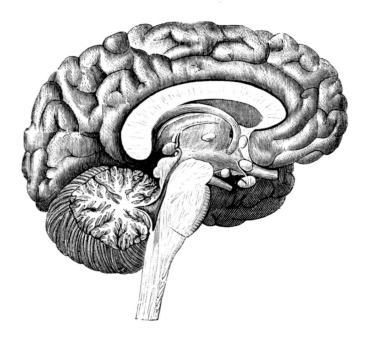

THE BRAIN IN YOUR HEAD.

This is what you automatically think of when you hear the word, "brain."

Your head brain has 86 billion neurons, the cells that process and transmit information. It's where synapses, electrical impulses, and hormones talk to each other, which is what allows for consciousness and awareness. Most importantly, it's what gives you the ability to identify patterns and make sense of the world.

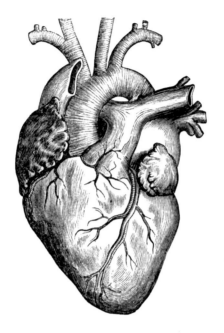

BRAIN #2

THE BRAIN IN YOUR HEART.

Yep, your heart has a brain, too.

Your heart houses more than 39 million neurons. It's not nearly as many as the head, but it fills this gap with generating the largest electromagnetic field in the body. The heart sends as many messages to the head brain as it receives. Researchers in the field of energy cardiology have discovered that your heart creates thinking hormones similar to the type created in the head brain.

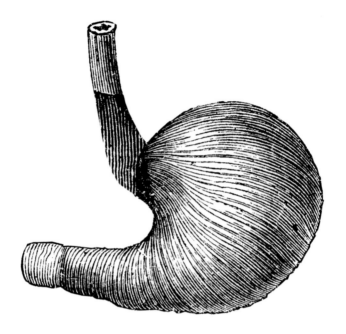

THE BRAIN IN YOUR GUT.

Your gut brain consists of two nerve centers called the myenteric and the submucosal, which have approximately 100 million neurons. This is more than the spinal cord.

The gut produces 70 percent of the hormone cortisol, which is released during stressful periods to regulate metabolism, control blood pressure, and assist with memory formulation. This gut brain is also responsible for processing information during sleep. 70 percent of serotonin—the neurotransmitter responsible for relaying signals across the brain to help you think clearer—is produced in the gut.

WHEN YOU LEARN
TO LISTEN TO ALL
THREE BRAINS,
SOMETHING
REMARKABLE
HAPPENS: YOU
FINALLY GO ALL IN
ON LIFE.

THE JOURNEY TO SHIFT...

SHIFT THE WORK

THIS ISN'T JUST A BOOK.

IT'S INTENDED TO
INSPIRE YOUR SOUL.

GRAB A PEN, TURN THE
PAGE, AND GET READY TO

DO THE
WORK!

PROLOGUE

DEATH IS DIFFERENT WHEN YOU FIND IT ON THE BATHROOM FLOOR.

DEATH IS DIFFERENT WHEN THE WALLS ARE SPLATTERED WITH BLOOD.

DEATH IS DIFFERENT WHEN THE SMELL OF ROT IS IN THE AIR.

And death is definitely different when the corpse before you is not only the woman who brought you into this world, but the scene reveals she violently struggled until her last breath.

Since seeing my mother's lifeless body sprawled on the floor, the fear of death has been a constant companion in my life, and the wish to go peacefully—with no regrets—has taken on the utmost meaning.

THE ENGAGEMENT CRISIS

"Put those fucking journals away," our companion said, looking up from his menu. My friends Yanik Silver, Joe Polish, and I had pulled out our pens and notebooks, intent on capturing the brilliant ideas of the man joining us that day for lunch.

Richard Saul Wurman is an icon of the digital age. As the founder of the TED conference in 1983, he foresaw how the convergence of technology, entertainment, and design would come to define the next generation of human advancement. Thanks to Yanik's ingenuity and courage, the three of us now had the extraordinary opportunity to pick this giant's brain.

Eighty years old at the time of our meeting, Wurman had already written close to eighty books on wildly different topics, the majority exploring how humans learn. His great insight was recognizing that industry leaders were not the only people capable of imparting knowledge. Wisdom, he believed, could come from the bottom up, from everyday people who mastered particular areas in life and work. Anyone, according to Wurman, has the potential to influence and shape the world.

In person, Wurman was stiff and distracted, as if he couldn't slow the many thoughts running through his mind. Then came his demand that we put our notebooks away.

"Writing notes, research shows, helps reinforce memory," Yanik defended.

"Life shows me that's bullshit," Wurman answered. *"Learning is remembering what you're interested in."*

We chewed our breadsticks slowly, digesting his words.

"Everything you need to know from this lunch, you'll remember," he explained. "You won't forget a thing you were interested in."

It's an idea that has informed his life and career. A person doesn't write ninety books on almost ninety different topics—or create an entity as multifaceted as the TED conference—without having a deep enthusiasm for all his subjects.

On the flight home, I reflected on the connection between interest and learning. It made perfect sense. Don't students excel in the subjects they find most interesting? I began wondering how it related to my own work.

OUR MISSION IS TO SHIFT THE WORK TO TRANSFORM THE WORLD.

Since 2001, we've been in the business of helping organizations—and individuals within them—strive, strengthen, and surge! Our successes, pathways, and proprietary methodologies have been forged in the beautiful struggle of helping more than six hundred businesses flourish.

At the time of the meeting with Wurman, my colleagues and I were struggling to wrap our heads around one particularly alarming statistic about the American workforce.

70% OF WORKERS

A) THINK THEY AREN'T GOOD AT THEIR JOBS, OR

B) AREN'T PASSIONATE ABOUT THEIR JOBS, OR

C) DON'T BELIEVE THEIR WORK IS ABOUT SOMETHING BIGGER THAN THEMSELVES.

In other words, these workers have "checked out." For seventeen years, this number has remained unchanged.[Intro:N1]

How can businesses expect to thrive if close to three-quarters of the workforce shows no passion—or even interest—for the work?

This led to a frightening realization, since the work world and the real world mirror one another.

If we aren't ALL IN at work, how can we expect to be ALL IN at home and in our personal lives?

Our lack of engagement at work follows us home.

I REPEAT: OUR LACK OF ENGAGEMENT AT WORK FOLLOWS US HOME.

Take out your pen and write out that last sentence; I want it to sink in.

WRITE IT OUT

....

....

Have you ever felt trapped at a company because of their lackluster mission or misaligned values?

When we leave work feeling depleted and hopeless, we don't return to our families and communities feeling empowered, looking for opportunities to innovate and create in our personal lives. Instead, the moment we step into the house, we put ourselves in front of the television to numb the pain with the mistaken belief that we lack a better option for how we spend our days.

"We the people" do whatever is easiest, which is maintaining the status quo and allowing inertia to take over.

We hate our politicians but can't be bothered to vote.

Who has the energy to get involved after a long day at work and taking care of the children?

We complain about high medical costs, yet we are the most obese nation on the planet.

Our schools are failing, inequality is growing, opioid abuse is destroying entire communities, and six million citizens are either in jail or on parole, yet a majority of our focus is reserved for TMZ, Tinder, and Facebook. Imagine the change we could bring if we held ourselves to a higher standard at work, trying to bring meaning to the one activity that makes up the bulk of our day.

Imagine the compassion we could unleash if we acted on our convictions and started to view apathy as the plague it is!

THE ENGAGEMENT CRISIS
HITS HOME

West Baltimore. April 2015. Riots break out after the death of Freddie Gray.

Businesses destroyed, hundreds arrested, the Maryland National Guard deployed. The hopelessness triggering this chaos is familiar to me. I grew up in East Baltimore, several miles away. Freddie Gray hailed from Sandtown-Winchester (West Baltimore), where unemployment rates sit at 50 percent. The other 50 percent that does work earns a median income of $24,000, well below the federal poverty level for a family of four.
Intro:N2

MASLOW'S HIERARCHY OF NEEDS

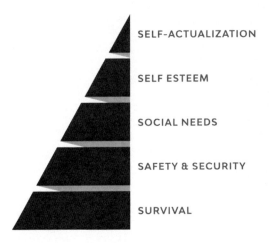

SELF-ACTUALIZATION

SELF ESTEEM

SOCIAL NEEDS

SAFETY & SECURITY

SURVIVAL

These people are on the bottom rung of Maslow's pyramid.

Every day is a fight to obtain the basic needs of food, clothing, and shelter.

Rising to the pyramid's second tier—to satisfy psychological needs like esteem and belongingness—is a pipe dream. The people in these communities struggle to find a purpose beyond survival.

Environments of despair have a clear mental effect on people, but they also—as this book will demonstrate—impact the body's physiology and biochemistry.

What if these workers had jobs that lit them up?

What if these workers enjoyed environments with equity, equality, and justice?

What if the companies they worked for could bring the best out of their employees?

What if the 70 percent of the workforce that feels disengaged succeeded in positively shifting their work?

CUE THE CHANGE...

The brains and bodies of these workers would change.

Workers would feel fulfilled, knowing they were making an impact at the place they spend a third of their life.

Employers would profit from a workforce that now went ALL IN on the job.

Productivity would rise, a more genial culture would emerge, and employees would spend time thinking about how to innovate and improve efficiency, instead of counting down the minutes until they could punch out.

Communities would benefit from citizens, fathers, mothers, and children (of working age) who now leave for work every day with a sense of purpose and come home feeling empowered to have a positive impact on their surroundings.

The possibility of creating nourishing work environments and connected communities doesn't apply only to the people of West Baltimore.

The entire country stands to gain if we can engage the workforce. In fact, the entire world does.

EVERYONE TO GO ALL-IN?

"Work isn't supposed to be fun."

So goes the typical response from earlier generations when asked about the engagement crisis.

At one time, this was true. Earlier generations had limited choices when it came to work. Skills, education, location, and family connections generally determined a person's job choice. If you lived in a coal-mining town, you were destined to become a miner. If your father was a banker, welcome to Wall Street. Middle-class workers with college degrees could expect a nine-to-five job at a company the person would call home for the next forty years, with every promotion and pay raise set according to a schedule. Nobody questioned these choices because people needed these jobs in order to cover basic needs—security, food, shelter—the bottom levels of Maslow's hierarchy of needs.

America has attained unprecedented levels of prosperity over the last thirty years.

This achievement gives people an extraordinary liberty to choose careers and jobs that light them up inside.

The time has come to capitalize on this newfound freedom.

WE DO NOT HAVE TO ACCEPT A 70 PERCENT DISENGAGED WORKFORCE AS A FACT OF LIFE. IT'S NO LONGER A HARD TRUTH ABOUT THE WORLD.

The newest generation to enter the workforce is seizing this freedom to change the culture of work. Millennials—people born between 1981 and 1996—are well educated, value time over money, and have a refreshing perspective on what truly matters in life.

At the end of the day, when millennials look themselves in the mirror and ask, "Was today worth it?" they expect a meaningful answer.

Happily bouncing from job to job, they seek what both excites them and matches their values.

30 percent of Millennials are emotionally and behaviorally connected to their job and company.

60 percent of Millennials are open to new job opportunities.

36 percent of Millennials will look for a job with a different organization in the next twelve months.[Intro:N3]

In the coming years, this trend will intensify. The Department of Labor estimates that today's students will have somewhere between ten and fourteen jobs by the time they're thirty-eight years old.

Millennials are why companies like Zappos and Google constantly make headlines for bringing positive changes to the work culture. They are in tune with the demands of these younger workers who expect purpose from work and the workplace.

They need to believe the company's mission aligns with their values.

It's no accident that these companies boast high levels of employee engagement either. In words and actions, they embody the idea that culture is the priority, not just a priority.

Despite these positive developments, change on a massive scale is still a long way off. Preventing it is the archaic notion that all decisions about work, like the preferences of millennials, are products of rational thinking. Science in the last several years is telling a vastly different story about how and why work engages us.

Recent discoveries in neuroscience reveal that at every moment, your body's neural network, which connects the brains in your head, heart, and gut, is sending signals about what it finds helpful or harmful.

The brain in your head controls and coordinates your actions and reactions, which allows you to learn, think, and store memories.

The brain in your heart beats to pump blood throughout your body, which transports nutrients and oxygen to cells.

The brain in your gut regulates your serotonin level, which dictates your mood and behavior.

In fact, there are countless ways these brains influence our decisions and behavior.

We will never solve the engagement crisis if we continue to ignore the brains in the heart and gut.

This book explores the emerging science around each of these brains, to help us move away from a work world that is designed solely to meet the needs of the brain in our head.

The three brains, when working together, don't lie. I know this from personal experience.

PERSONAL

Twenty-three.

That's the number of funerals I attended before my twenty-third birthday. The last three deaths hit me the hardest, coming at what should've been an exciting time in my life.

The HBO television show *The Wire* is the best comparison to the neighborhood in Baltimore where I was raised.

Violence and unemployment were rampant.

My high school was the worst in the state of Maryland with a graduation rate of 23 percent.

A tough upbringing. Yet, I succeeded in making it out. I managed to get accepted to Johns Hopkins University. Right after graduation, I was offered a job at the prestigious Andersen Consulting (now Accenture). Accepting this job would be the bow on the new life I had created for myself. Oddly, at this moment, I found myself stuck in place, unable to decide on my next move.

First came the death of my mentor.

During college, I was always starting different businesses, and one of them was a house-painting company. By the business's third year, I had more than a hundred people working for me. One day, some disgruntled workers trashed a mansion they were supposed to paint.

The owner, Stan Burns, insisted that I would personally repair the damages, even though I didn't know how to paint. I was scheduled to leave Baltimore that week to begin work at Andersen.

He demanded I take responsibility.

I complied, delaying the start date of my new job. Over the next few weeks, as I worked on his house, this former bank executive took me under his wing and shared his philosophies on life and business.

"You've got to love who you work with and love who you do the work for. That's the kind of passion that breeds success," he told me.

This was a radical idea for a kid like me with a blue-collar upbringing. I'd always thought of work simply as a way to pay the bills. He suggested that someone with my background might not be happy in a suit-and-tie environment like Andersen. His words hit me because I wasn't excited to say yes to the people at Andersen, but I wasn't prepared to say no either.

I settled for pushing off my start date. Meanwhile, some buddies and I started a different business. It involved being on the road seven days a week, a different city each day.

Upon returning to Baltimore after one trip, I received a call from Stan's daughter with news that he had passed unexpectedly.

Three weeks later, my cousin was killed in a drunk-driving accident.

Three weeks after that, my mother died. I'd be lying if I told you that her death didn't come with a huge sense of relief.

For twenty-three years, I was my mother's caretaker.

She called whenever she was sick, hungry, or troubled. She developed diabetes when pregnant with me and struggled with the disease for most of her life. She smoked and drank. She didn't watch her diet, and I remember judging her (as a child) for the careless way she treated her body. My grandparents helped raise me.

At age five, I went to live with my mother for the first time, and the two of us lived on food stamps. Opening the refrigerator and seeing bare shelves is still the worst feeling I've ever had in my life. Since, my wife has had to get used to me buying too many groceries at the store.

This is not to say that my mother and I didn't have a special bond. She was an aspiring actress, and at night, she'd make me read lines with her. We'd dance together across the living room floor. I was an only child, and I felt responsible for her physical and emotional health, although it came with resentment and, eventually, contempt.

When I turned thirteen, I finally built up the courage to move out of her house. We didn't speak for almost two years, until she got sick again.

Suddenly, I was in the hospital and at her side.

During my college years, she was in and out of hospitals, undergoing pancreas and kidney transplants. She became legally blind, and a broken hip left her in a wheelchair. Every Thanksgiving and Christmas was spent together in a hospital. She demanded a lot even when she wasn't in the hospital. In the middle of the night, she'd call asking for money.

College was hard enough without these distractions. Coming from the worst high school in the city, I was ill-equipped—emotionally and academically—to deal with the academics and my commitment to the football team.

My mother's last hospitalization was different. It was a heartattack, her first one. A full recovery was expected, and the doctor suggested it would be fine for me to head to New York City for a planned business trip. Three days later, I was speaking in front of a hundred people when my telephone began vibrating. I looked down and saw it was my mother. Given her condition, I decided to take the call. She was hysterical. I couldn't get a word in.

It was the same call I'd received a hundred times before, so I hung up on her and went on with my talk.

Several days after returning to Baltimore, I went to retrieve my checkbook at my father's house so I could pay my mother's rent. His phone rang. Oddly, I decided to answer.

It was the Baltimore City Police calling to tell him my mother was found dead.

A cop, coroner, and police administrator greeted me at her apartment door. Death was not new to me. I'd seen people die in front of my eyes. It was always a peaceful scene, in a hospital room or at home in a bed surrounded by loved ones.

The lifeless body of my mother within blood-stained walls was not a reflection of peace. The conclusion of a rough stretch, to say the least.

Everybody faces trials and hardships.
Everybody has a story.
Everybody chooses a path.

The purpose of sharing the difficulties from my life isn't to prove that I've suffered greater disadvantages or heartbreak.

It's to demonstrate how we hold the power to shift our perspective and make different choices, which is exactly what I set out to do.

"The way we do anything is the way we do everything."
MARTHA BECK

YOUR PATH

When you get clear on the path that led you to where you are now, you can get clear on where you want to go.

WHAT ARE THE KEY EVENTS THAT LED YOU TO WHERE YOU ARE NOW?

....

....

....

....

....

....

....

....

WHAT IS YOUR PATH FORWARD?

....

....

....

....

....

....

....

....

SHIFT THE WORK

TRANSFORM THE WORLD

I'd spent the first twenty-three years of my life beating myself up for feeling I was never doing enough for my mother.

As I embarked on this next phase of life, I was determined to do a better job of loving myself. This new mindset provided me with the courage to follow my dharma, my path.

Instead of pursuing the obviously safe, lucrative, and distinguished position at Andersen Consulting, I decided I'd do something challenging and momentous with my life. So I opened my own consulting firm.

Let's be honest: launching a consulting firm at the age of twenty-three is foolish.

The job of a consultant is to perform brain surgery on a company.

Other than starting some businesses during college, I knew nothing. Opening a lemonade stand would have been easier.

My mom died on July 10, 2000, and we started entreQuest, which has since transformed to SHIFT, on November 1 of that same year, meaning it took less than four months to forge ahead down this new path.

If my mother hadn't passed away, I would never have turned down the job with Andersen, completed a triathlon with my business partner, started a business, established a foundation to help Baltimore City youth, or spent my mornings writing about what I was grateful for.

This last activity ended up being the most impactful one. It has shaped all the other opportunities. Change your language; change your life.

Writing in a journal every morning prevented me from seeing terrible episodes as things that were happening to me. Instead, I saw even misfortunate episodes as things that were happening for me. More on that later.

A shift in perspective can transform a life. At the time I didn't appreciate the biological component to my change in outlook. That recognition came years later, and I learned it the hard way.

> "You are not a drop in the ocean.
> You are the entire ocean in a drop."
> **RUMI**

Grow Regardless, my first book—a *New York Times* best seller—was written to warn entrepreneurs that nobody is going to come save their organizations, not the government nor their industry.

The United States was founded on the idea of individual freedom, which is as much a sociological statement as a political one. It's the responsibility of executives and leaders to be culturally driven and to build organizations that matter. They'll succeed once they understand how the culture they create has the power to impact the engagement, performance, and productivity of their workers.

The book and its message affected tens of thousands of companies. We received notes and gifts of thanks from hundreds of organizations that revitalized their culture with the help of our tools and techniques.

The problem: it takes two to tango.

RIG THE WORLD IN YOUR FAVOR

Changing your perspective to an attitude of gratitude can lead to endless possibilities to positively change your world.

WRITE DOWN A NEGATIVE EVENT OR SITUATION THAT RECENTLY HAPPENED TO YOU.

NOW, REWRITE THAT SAME EVENT AS IF IT WERE A POSITIVE, UNFAIR ADVANTAGE FOR YOU.

The guidance we provided, we soon came to realize, is only part of the answer. Richard Saul Wurman's message suggested that change comes not from the top— the executive structure—but from the workers who take an interest in their jobs.

When the workers aren't ALL IN, according to Wurman's thinking, it's because they aren't enthused about either the work, their manager, the company, or its mission. To widen our impact, we need to reach the people punching the clock every day.

Tony Hsieh, the CEO of Zappos, wrote in his superb book, *Delivering Happiness*, "In a poker room at a casino, there are usually many different choices of tables. Each table has different stakes, different players, and different dynamics that change as the players come and go, and as players get excited, upset, or tired. I learned that the most important decision I could make was which table to sit at."

The table we are sitting at now is focused on shifting the workforce from 70 percent disengaged to 70 percent engaged. We're figuring out how to light up these workers, from the inside out.

Motivational self-help guides that preach positive thinking have helped millions of people but quickly become outdated with advancements in neuroscience. Behavioral economics, evolutionary psychology, organizational development, and sociology have taught us more in the last three years about the body's decision-making process than the preceding three thousand years combined. We're finding cures to cancer and fighting diseases through immunotherapy, yet employers and employees are slow to embrace the science about how we spend our time at work?

Since 2001, SHIFT has been working to solve this engagement crisis. The following pages deliver stories, anecdotal evidence, science, statistics, and strategies aimed at helping employees—at all levels—take responsibility to emerge from the fog of dissatisfaction. Change of this magnitude will not happen on its own. The eight to twelve hours we spend every day in a professional environment need to count, so we can go home feeling it was worth it.

In your hands is a manifesto. This is a Jerry Maguire moment. I may lose some clients along the way, and possibly even some readers, but I sincerely hope this doesn't include you. If some people think I'm taking an overly optimistic view of how workers will force companies to change, then they don't understand the severity of the crisis. They also don't grasp this unique historical opportunity to change the world together by changing the way we work.

THIS IS THE MOMENT WE SAY ENOUGH IS ENOUGH, AND WE BEGIN TO LISTEN TO WHAT OUR THREE BRAINS ARE TELLING US ABOUT THE WORK WE DO.

It's time to shift our thinking from dogma to data, from story to science, as we shift to purposeful engagement with the work world. When we care for ourselves enough to transform our work, we take the crucial step in beginning to properly care for others and our planet.

THE SINGLE GREATEST LEVER TO UNLOCKING HUMAN POTENTIAL IS A MORE ENGAGED WORKFORCE.

BETTER YOU.
BETTER US.
BETTER ALL.

"Follow your bliss. If you do follow your bliss, you put yourself on a kind of track that has been there all the while waiting for you, and the life you ought to be living is the one you are living. When you can see that, you begin to meet people who are in the field of your bliss, and they open the doors to you. I say, follow your bliss and don't be afraid, and doors will open where you didn't know they were going to be. If you follow your bliss, doors will open for you that wouldn't have opened for anyone else."

JOSEPH CAMPBELL

THE COMPOUND EFFECT

The first step toward positive change is *belief*. Belief that the single greatest lever to unlocking human potential is a more engaged workforce.

WHAT IS ONE CHANGE YOU CAN MAKE NOW TO BE MORE ENGAGED AND HAPPIER AT WORK?

CHAPTER 1

WORK ON

PURPOSE

"You have a right to perform your prescribed duties, but you are not entitled to the fruits of your actions. Never consider yourself to be the cause of the results of your activities, nor be attached to inaction."

BHAGAVAD GITA

OPPORTUNITY COST

Make a list of five work responsibilities or domestic chores you'll have to do once you get up from reading this book.

1. ....

2. ....

3. ....

4. ....

5. ....

SHIFT THE WORK

UP THE MOUNTAIN

It's 2013, and I'm flying high. Erica, my wife, delivers our second child, a baby boy to join our daughter, Ellie. A week later, I'm able to cross an item off my father's bucket list when I take him down to New Orleans to watch the Baltimore Ravens play in the Super Bowl. The next day, my first book, *Grow Regardless,* hits number one on the Barnes & Noble best-seller list. The next week, we learn it has cracked the *New York Times* list. It's obvious that the book is reaching people. *Fox News* and *Bloomberg* bring me on the air to discuss *my* ideas. I'm invited to conferences, book parties, and readings. In my mind, the book has turned me into a minor celebrity. To top this off, I receive a call from the CEO of a Berkshire Hathaway subsidiary who, aftering reading the book, now wants my company's help.

I'm riding high. Life is good.

Except that it's not. I'm buried under clouds, instead of riding high on them. Acid reflux is keeping our son up at night. Erica and the baby haven't slept for two months. At night, all I contribute are complaints about the endless crying. During the day, whenever it's clear that my wife or kids need me to chip in, I slip away, grumbling about a busy work schedule and book commitments. Time with the family begins to feel like an obligation, and I hear it in the language I use to describe basic parenting responsibilities.

I have to do bath time.
I have to feed the kids.
I have to spend time with them.

One night, after a full day of interviews for the book, I'm lying in bed listening to my son scream his head off in the nursery. My legs feel heavy, and I can't muster the energy to get out of bed and check on him so Erica can continue sleeping. Instead, I keep my eyes closed, hoping my wife will wake up and do the job of comforting our son.

Meanwhile, the entire time, I'm thinking that I'm the worst dad ever and not such a great husband. The next morning, my wife confirms this epiphany, gently telling me that I'm an amazing father to only one of our children. It's time, she suggested, to start showing equal patience and kindness toward my son.

Her unique ability to call me on my bullshit is why I fell in love with her. I may have felt as if I was on top of the world, but Erica could see that I wasn't reaching my potential as a father or husband. I'd lost sight of the fact that we had two healthy children and a good life. A child keeping me up at night was nothing more than a minor frustration.

Major hardship, on the other hand, was something both my parents knew well, yet they never lost sight of the gifts life gave them.

ATTITUDE OF

GRATITUDE

It wasn't part of my father's life plan to start a family right after graduating high school. He had a tough upbringing, with an absent father and a mother who married five times (twice to the same man). Finally out of the house, my father spent most of his days doing things he shouldn't have been doing. At twenty, he met my mom, who was hitchhiking across the country. Months later, she was pregnant. My father didn't respond to the news by sticking his thumb out and hightailing it out of town, even though it would've been natural to treat the unborn child with the same indifference shown to him when he was a boy. Instead, the way he tells it, something inside him shifted. Fortunately for me—as the child born from this union—my father burned with the sudden desire to become a great father.

It was a desire that followed him into his second marriage to my stepmother, Rose—a warm, sincere, and selfless woman who had a way of showing me that there is always an option to take the high road and truly be a better person.

Sweeping floors was the only job my dad could get without a college degree. The fact that it was mundane, tedious work didn't discourage him. From his first day at the new job, he shifted the work and went ALL IN, appreciating it as an opportunity to provide a better life for his family. Day after day, he clocked in with an attitude of "I get to be here," as opposed to, "I have to be here." An attitude of gratitude accompanied him from home to work and back. This sense of purpose, passion, and perspective allowed him to come home as a better father, husband, and neighbor. Within several years, he was driving the forklift outside, sweating in the summer and shivering in the winter. Seventeen years later, he was the senior vice president and partner. He's now worked at the same company for close to forty years, helping grow a small business into one with several thousand employees.

No matter how bad we had it, my mother always insisted we lift our heads and open our eyes, so we could see that there was a larger world around us with greater concerns. Local pantries would deliver us food, and my mother would give some of it away to our even hungrier neighbors. She refused to see the pains and hardships she suffered as the world's punishment. Instead, she saw them as gifts. They reminded her that life is short and precious, and that every minute counts. She was my inspiration.

My mother taught me, among many things, the law of familiarity: the closer we get to someone or something, the more we take it for granted. We begin to think of ourselves as objects in relation to the other person or thing. We feel as if that person or job is a permanent feature in our life, and there is nothing we can do to control it.

For twenty-three years, the role I assumed as the rescuer defined my relationship with my mother. After she died, guilt paralyzed me. I walked around convinced there was something I could've done to save her. I couldn't forgive myself, and I had trouble moving ahead with my life. I was unsure of whether I should take the job at Andersen Consulting or strike out on my own. It was only when I stopped playing the victim and began focusing on my parents' lessons of gratitude and familiarity that my mind and body began to feel free.

It started with the simplest of exercises. Every morning, I'd take ten minutes to write in a journal the small and large things in life that made me feel grateful.

> SLOWLY, I ARRIVED AT A PLACE WHERE I STOPPED BELIEVING THINGS WERE HAPPENING *TO* ME, AND INSTEAD UNDERSTOOD THAT THEY WERE HAPPENING *FOR* ME.

This small change in syntax changed my perspective. I took ownership of the obstacles that came my way. I was now a subject in my own life, a person who was in complete control of my destiny. This perspective challenged me to turn down the job at Andersen and to chart my future by starting my own company.

"People only see what they are prepared to see."
RALPH WALDO EMERSON

DO THE WORK

VICTIM TO VICTOR

WHERE HAVE YOU BEEN PLAYING THE VICTIM, SHIFTING THE BLAME INSTEAD OF TAKING RESPONSIBILITY? NAME IT.

SHIFT THE WORK

THE FIVE MINUTE JOURNAL is a great tool to enhance your optimism and happiness and to build stronger relationships. At the start of the day, you are asked to list three things you are grateful for, three things that would make the day great, and a final affirmation of how you see yourself. Then, at the end of the day, you list three amazing things that happened, and one way the day could have been better. Articulating your dreams and desires has the power to make them a reality.

Gandhi said, "Your beliefs become your thoughts; your thoughts become your words; your words become your actions; your actions become your habits; your habits become your values; your values become your destiny."

The attitude of gratitude still guides me through my days. It's why I turn off my phone at night and spend quality time with my wife and kids, or why at work I strive to treat our smallest clients the same as our largest accounts, grateful and humbled that they selected my company over all the others. Still, there are moments when I take things for granted. It's what happened during the first half of 2013 when I was riding high on the success of my book.

ATTITUDE OF GRATITUDE

WRITE DOWN THE TWENTY-FIVE THINGS YOU ARE MOST GRATEFUL FOR.

1.
2.
3.
4.
5.
6.
7.
8.
9.
10.
11.
12.

13. ....

14. ....

15. ....

16. ....

17. ....

18. ....

19. ....

20. ....

21. ....

22. ....

23. ....

24. ....

25. ....

"As we express our gratitude, we must never forget that the highest appreciation is not to utter words, but to live by them."

JOHN F. KENNEDY

THE FALL

People think reaching the summit is the most difficult part of climbing a mountain. As someone who climbed Kilimanjaro, I can attest that the climb down is much harder.

My descent started in the second half of 2013, on the heels of my wife calling me out for not being my best self. The Berkshire Hathaway company that contacted us after the CEO read my book was lining up to become SHIFT's big break. For eighty-seven days, we worked our tails off putting together proposals and action plans. Then, the day before we submitted our work, the CEO, our main point of contact, was removed from his job. We soon learned that nobody else in the company knew anything about this project, rendering our work irrelevant overnight.

Weeks later, two of our teammates were in a horrific car accident. The team maintained a vigil at the hospital as our coworkers fought life-threatening injuries. It was touch and go for the next month, but they pulled through.

Shortly after, I took my own trip to the emergency room with severe stomach pain. The intensity of the pain was such that they connected me to a morphine drip. The doctor found a massive infection and advised me to stop eating processed foods. When I told him I don't eat processed foods, the doctor asked if I'm dealing with any stress. "Where do I start?" I answer.

I'd ridden the roller coaster all the way down, and I felt as if I lacked the mental strength to scale the summit again.

Here it is again: change your language; change your life.

It was time for a shift to gain a healthier perspective by embracing an attitude of gratitude. Things weren't happening *to* me. They were happening *for* me. Every instance of misfortune is a gift, a reminder of life's preciousness and the importance of relationships.

I don't *have* to be with the people I love. I *get* to be with the people I love. I *get* to bathe my children. I get to feed my children. These obligations toward loved ones are not a punishment but opportunities to hug them a little tighter and deepen the bond. Then, I took the same approach with my work obligations. I *get* to answer this email. I *get* to have a meeting with a prospective client. This small change allowed me to reconnect to my priorities, passion, and purpose. It empowered me to shift my work.

REVISIT THE EXERCISE AT THE START OF THIS CHAPTER, AND REWRITE EACH LISTED ITEM, ADDING "I GET TO" BEFORE EACH STATEMENT.

THIS IS YOUR

BODY TALKING

Changing language to change perspective is not motivational rhetoric, a placebo effect, or a modern take on Norman Vincent Peale's *The Power of Positive Thinking*. Something physical occurs when we change our language. Consider this study out of Stanford University: high levels of serotonin, the neurotransmitter that regulates feelings of happiness, are released from the gut when people express gratitude toward one another while performing acts of generosity. The language we utter, in other words, has a measurable biological impact. Our perspective shifts.

Studies have shown that the more we practice gratitude, whether through journal writing or sending out thank-you cards, the more thankful we will feel months later. This shows up on brain scans. It's as if brains practicing gratitude will finally adopt it as a permanent mindset.[CH1:N1]

None of this is surprising. The stomach infection I developed was a physical response to the language I was using to interpret the negative events in my life. Stress was causing me to lose sight of my priorities, passion, and purpose. Meanwhile, my head, heart, and gut were sending me warnings that they weren't properly aligned, but I was ignoring them. Finally, they sent me a loud signal in the form of an infection that became impossible to dismiss.

In *Why Zebras Don't Get Ulcers*, Robert Sapolsky outlines how stress can lead to not only poor decision-making but eventually disease and disability. Science, biology, and math are what we're missing in the workplace.[CH1:N2]

If ignoring the brains in our body can put us in danger, then listening to the signals they send can allow us to achieve greatness.

In recent years, science has painted a fuller picture of the control we have over the three brains in our body—head, heart, and gut—and our ability to use them as guides.

For example, study after study proves the tremendous malleability of the brain in the head, called neuroplasticity. The notion that at a certain age the brain in the head becomes fixed and unable to grow new cells is antiquated. Amino acids and caffeine in green tea improve brain function. Restful sleep clears waste from the brain and improves concentration and memory. Eating foods rich with omega fatty acids like pecans, pistachios, leafy vegetables, and olive oil helps brain development and the prevention of attention disorders and brain diseases like Alzheimer's. Mindfulness has proven the ability to teach a brain how to deeply focus on the present moment. Malcolm Gladwell and Anders Ericsson have written books on the brain's elasticity and ability to gain new skills through practice and diligence. What's clear is that the brain in the head has the power to develop new outlooks and patterns of behavior late into life, meaning we aren't at the mercy of our genes.

IF YOU ARE NOT WILLING TO LEARN, NO ONE CAN HELP YOU. IF YOU ARE DETERMINED TO LEARN, NO ONE CAN STOP YOU.

Great rewards await you when you provide all three brains with the materials and actions needed to thrive!

Did you know that the brain in your head is not the ultimate orchestrator of bodily function? Turns out there's a neural network connecting the brain in the head with the brains in the heart and gut. Your somatic nervous system is what's responsible for the conscious movements of your body's muscles. Your autonomic nervous system is what moves parts of the body without conscious control. In other words, you don't have to instruct your heart to beat, or your gut to digest food. They do it on their own with the help of their distinct brains.

Let's explore what other powers these brains in your gut and heart possess!

Through the head, heart, and gut, our neural network sends signals to help place us in the right environments and relationships, so we can be ALL IN on life. Are you listening to your body and aligning your life according to the signals you hear?

Understanding the function of the brain in the gut could have helped me decipher the message it was trying to send when the infection developed in my stomach. The gut is the spiritual center of our body. When we use words like intuition, inspiration, and sixth sense, we are talking about sensations that emanate from the gut. It's the place people point to when mentioning their soul and where the connection between people is made. It the part of our body that most represents our shared humanity. Every person, rich or poor, has a gut that processes the energy required to sustain life. It takes what it needs and gives back to the world what it doesn't, so it can create more energy for other beings. It's the reminder that at the core we are all the same. We all come from the same DNA. The same energy. The same minute particles of matter. We are all stardust.

Since the gut is about purpose and shared humanity, it's no wonder I developed an infection in this area when I'd clearly lost touch with my ideals.

SCIENCE AND

THE WORKER

In the last fifteen years, SHIFT has worked with six hundred companies—employing five hundred thousand people—to create positive workplace cultures. We've definitely had our share of successes on the management side of the equation, yet we still see employees struggle to overcome disengagement, distrust, and distraction.

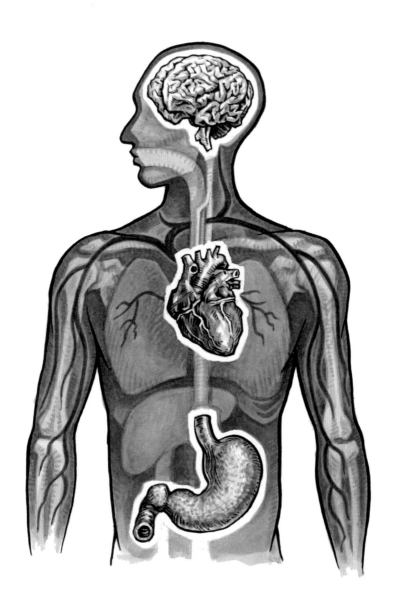

Money is no longer the chief fixation for workers. They want to know that work is worth their time. These workers will flip burgers for a living if they're doing it with people they love and for people who they believe need the product.[CH1:N3]

THE WORKPLACE FACTORS THAT MATTER MOST TO EMPLOYEES

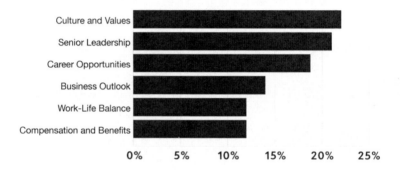

Note: Horizontal axis shows the relative contribution of each factor to overall satisfaction, where all factors sum to one.
Source: Glassdoor Economic Research (glassdoor.com/research)

The highest-performing teams in sports are the ones where the players are connected to their heads, hearts, and guts. The teammates show up with passion, energy, focus, and goals. It allows them to play selflessly. Interestingly, when assessing a sport's greatest heros, failure of a team to win a championship is counted as a strike against a single player's greatness.

Businesses struggle to create environments in sync with the bodies of their employees because they're disconnected from the workers' real needs. Business speaks in the straightforward language of productivity, efficiency, and profits. These goals can still matter in a world where workers don't hate their bosses and jobs. In fact, workers will happily go ALL IN and shift the work if their minds and bodies are healthy.

Work cultures will only change if the bulk of the initiative comes from the workers. The leadership of companies can try to divine what the typical worker wants out of work, but, at the end of the day, only workers can hear the signals coming from the brains in their own heads, hearts, and guts. It reminds me of a line from the movie *Hitch*. Will Smith's character is teaching his protégé how to pull off the perfect first kiss. He says, "The secret to a kiss is to go 90 percent of the way and then hold." Kevin James's character asks, "For how long?" To which the Will Smith character answers, "As long as it takes for her to come the other ten."

Change requires people to listen to their bodies' plea to take their careers down a different path. If they collectively act on these signals, the management will have no choice but to go the other 10 percent of the way.

Unfortunately, workers lack the tools to regulate the neurotransmitters wreaking havoc with their moods, perspectives, and even physical health. Soldiers, for example, are trained to exhibit courage under fire, which really means maintaining healthy levels of serotonin so clear thinking can take place. Workers, on the other hand, don't receive this type of training. In a world of deadlines, negative feedback, and difficult colleagues, a worker's stress is off the charts. Overwhelmed, the person feels like his or her back is against the wall. Things are happening *to* the worker. The worker *has* to perform certain tasks. The worker has become an object with no control of the situation and is powerless to make changes.

As human beings, we crave the security and predictability of homeostasis, even when it's not comfortable or advantageous. We ignore the signals and don't even consider whether regulation is possible.

Fortunately, there is another side to humans. The other part of our DNA yearns for variety, adventure, and richer levels of experience! The current science tells us that our bodies are constantly trying to connect us with this other aspect of our makeup. If we have any hope of solving the engagement crisis, we first need to understand this underlying drive for a higher purpose and richer life.

BODY TALK

WHAT IS ONE THING YOUR BODY IS SIGNALING TO YOU THAT NEEDS TO CHANGE IN YOUR CURRENT WORK ENVIRONMENT? WRITE IT DOWN.

WHAT IS ONE POTENTIAL SOLUTION TO IMPLEMENT THE CHANGE YOU SEEK?

CHAPTER 2

EVOLUTIONARY ENGAGEMENT

DO THE MATH

The talent gap as it stands now: six million openings for jobs, with six and a half million unemployed workers.[CH2:N1] Even if we managed to fill only a quarter of those job openings, we'd boost GDP and improve the lives of millions of people. The math seems rather simple, yet we lack the political will to help these people. Instead, we promise to bring back the jobs of dying industries that rely on ancient, discarded technologies. We've told ourselves that it's not possible to shift these people into new careers. "They don't have the skills." "The companies lack imaginations." "They can't entertain hiring someone without a traditional degree from a four-year college." Meanwhile, these people don't have the funds to go back to school. All of this means it's easier for politicians to promise a yesterday that's not coming back than to ask constituents to face reality.

On a recent trip to Jerusalem, I was fortunate to be hosted by top tech blogger and advisor Hillel Fuld, while on a tour of the Israeli tech scene.

SIDENOTE:
Hillel posts a weekly vlog that will blow your mind: hillelfuld.com

There, I had the honor of visiting the offices of Hometalk, the world's leading community for DIY home and garden projects. Not only do most of the company's computer programmers come from the ultra-Orthodox community, but the majority of them are women. These are people who lack a traditional education. They are coming from a community that until recently looked askance on jobs outside their communities. People consider this a revolution and talk about the possibility of raising an entire community out of poverty. Hometalk is seeing the monetary and societal benefits of going after this untapped labor market. But what was most striking when visiting these offices was the attitude I saw on the faces of the workers. It was clear that they felt they were part of something big, a major shift in the way their CEO thinks about the workplace.

It will take tens of thousands of companies to shift like Hometalk before we start seeing the end of the engagement crisis. Even though it's a recent development, it's one whose roots have been firmly planted over the years—close to three million, to be exact.

THE FACE OF CHANGE

If you're reading this book in a crowd of other people, take a moment to look at their faces.

DO YOU SEE JOY? PURPOSE? ENGAGEMENT? WHAT ARE THEY SEEING IN YOUR FACE THIS VERY MOMENT? WHAT DO YOU WISH THEY SAW?

WRITE IT DOWN.

SHIFT THE WORK

THE WORKPLACE

More than 2.8 million years ago, humans started using tools, which had a major impact on their diet. Tools gave them greater access to protein, fat, nutrients, and calories. Brains got bigger. Guts shrank in size. These changes altered the social order. Some people excelled at using the tools. Women with greater access to food had an easier time raising offspring. Humans developed the concept of jobs and responsibilities. The hierarchical system was introduced.

It was 28,000 years ago that the brain began to shrink as population density increased. Complex societies, or tribes, emerged. Brains were smaller because people didn't need the same smarts to keep themselves alive. They could now rely on their neighbors to help keep them safe.[CH2:N2] The change in size doesn't mean humans became dumber. The brain and its wiring simply developed into a more efficient muscle.

Only 280 years ago, the industrial revolution changed the world of commerce. Industry became occupied with the question

of how to manufacture products on a mass scale. During this period, hierarchies took even greater shape within businesses. Workers had specific roles and responsibilities within an organization. Conformity ruled the day. Compensation and job security became mechanisms to enforce compliance. This established a framework where society was shaped around business, and not the other way around. The way we educate our children today, for example, is based on the educational model from this period, which attempted to create large groups of compliant factory workers with skills to execute limited tasks.

For the first hundred years after the industrial revolution, organizations maintained a control and compliance-based hierarchical system. Many workers were automatically placed at the bottom of the pyramid with no chance of advancement. Companies didn't look favorably on women or minorities. This stringency—discrimination, really—prevented them from tapping major pools of talent. Up until twenty-eight years ago, companies mostly maintained these reactionary attitudes. Any progress toward a more inclusive and less hierarchical environment was incremental.

Creative and information economies emerged in the decade before the internet was commercialized. Conformity began to succumb to the power of these new market forces. Workers in this new age had less interest and tolerance for working in such homogenous and discriminatory environments, and the pace of progress sped up.

Businesses slowly grasped that satisfied workers meant greater production and greater profits. Companies realized that if they wanted the next million-dollar idea, they needed to get the workers' creative juices flowing. Conformity wasn't going to cut it any longer. Industry norms tried—and largely failed, but tried, nevertheless—to reflect the wants, needs, and desires of its workers. Workers could dress however they wanted, and the walls that used to divide people by position and status were knocked down or turned to glass. Ping Pong and Foosball tables were placed around the office. Break rooms were stocked with free drinks, energy bars, and fruit. Stock options became a centerpiece of employees' compensation.

All these developments reflected the changing attitude of workers. Generation Xers valued time over money, a novel idea for its time. Put simply, they wanted to enjoy their time at work. The scales began to tip, and, increasingly, businesses started to become more of a reflection of society.

"You can only become truly accomplished at something you love. Don't make money your goal. Instead pursue the things you love doing and then do them so well that people can't take their eyes off of you."
MAYA ANGELOU

THE FORMATIVE YEARS

What did you LOVE to do in grade school that made you avoid doing homework? Think about the hobby you stayed up doing all night.

WRITE IT DOWN WITH AS MUCH DETAIL AND COLOR AS POSSIBLE.

HAVE YOU EVER THOUGHT ABOUT EARNING A LIVING FROM IT?

SHIFT THE WORK

This trend will deepen in the coming years as millennials, a group that thinks about work in a completely unique way, join the workforce in greater numbers. In ten years, they will make up 75 percent of the working population. Many people dismiss the millennial generation's attitudes toward work as nothing more than an unwarranted sense of entitlement.

THIS ASSESSMENT MISSES SOMETHING CRUCIAL: IT'S ALL ABOUT BIOLOGY.

The bodies of millennials expect different things from work because their bodies know they are living in a time of tremendous prosperity. In the last seventeen years, technological and medical advances have lengthened life expectancy and provided these workers with comfortable lives. Food and shelter are taken for granted. Meanwhile, thanks to mobile devices, the brains in their heads are processing a tremendous amount of information. The brains in their hearts experience a myriad of feelings as they encounter scenes from across the world. And the brains in their guts know what it means to live for a higher purpose.

Millennials have greater needs than mere survival. What their bodies want is to find engaging, impactful work, no matter the organization. When they lose that feeling at one job, one of two things happens: some will leave their jobs in search of more fulfilling work, while others will stay put, squandering this historically unique opportunity to take advantage of this age of prosperity by finding meaning.

The importance of time and meaning over money is so great that many of them are willing to work as baristas and live in their parents' basements rather than work long hours at a job they don't love or believe in. They have no need to own a home or car or to maintain a savings plan. Whether or not the work has value is their primary concern, and they're willing to wait for the right situation. It's admirable that they can display such self-determination and patience, to be honest. Fortunately for them, the economy is starting to recognize these innovative values. Renting, leasing, and even sharing have never been easier. Working remotely is commonplace.

The emergence of companies like Lyft and Upwork, and the entire gig economy, is an indication that we have entered a new era. Successfully tapping into the evolutionary changes undergone by millennials, these companies are keen to provide the users of their services with more time to focus on the activities that will light them up, and make them feel more engaged in every moment of the day. On the flip side, the freelancers who work for these companies are usually people who want control over their time and don't want to be tethered to the traditional demands of a company.

> *"It is our choices... that show what we truly are, far more than our abilities."*
> **J. K. ROWLING**
> *HARRY POTTER AND THE CHAMBER OF SECRETS*

DO THE WORK

MILLENNIAL MEANING

WOULD YOU RATHER:

a) work long hours at a well-paying job you hate, or
b) work shorter hours at a low-paying job you love?

Note: If you were born after 1981 (the official beginning of the millennial generation), this should be an easy question to answer.

A REVOLUTION?

Modern commerce has brought phenomenal progress for humanity. Keeping the ambitious Sustainable Development Goals in mind, the rate of poverty reduction has scaled tremendously, cutting the 1990 poverty rate down 50 percent by 2010, five years ahead of the 2015 goal schedule.[CH2:N3] In this country, however, we've hit a standstill in terms of growth.

As a country, we're convinced of the idea that we're number one. According to a select host of metrics, the United States is indeed the best. This has led us to a state of complacency. It's hard to go ALL IN when you think you don't need to. Meanwhile, our failure to go ALL IN means we're neglecting the many areas where we're not number one, where we're falling a little bit short.

Even a quick look at society exposes how we waste our tremendous prosperity. The staggering amount of data collected on the United States' population in the areas of health, economy, and incarceration serve as sobering reminders and catalysts for action.

According to a 2013 Gallup poll of 1,039 Americans, obesity ranks as the nation's most urgent health ailment, before even cancer and heart disease. For every three people in the United States, one is obese.[CH2:N4] How did this creep up on us to become an epidemic?

Our debt-to-economic output (GDP) ratio sits at a shocking 104 percent for the third quarter of 2017. According to the World Bank, we've blown past the safe zone "tipping point" at 77 percent, which means increased interest rates and slower economic growth.[CH2:N5]

U.S. INCARCERATIONS

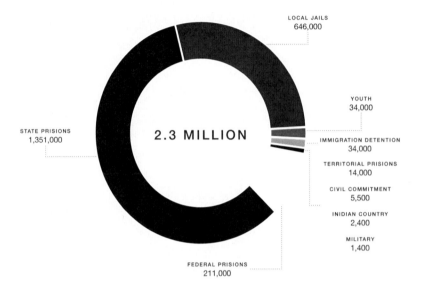

LOCAL JAILS
646,000

STATE PRISIONS
1,351,000

2.3 MILLION

YOUTH
34,000

IMMIGRATION DETENTION
34,000

TERRITORIAL PRISIONS
14,000

CIVIL COMMITMENT
5,500

INIDIAN COUNTRY
2,400

MILITARY
1,400

FEDERAL PRISIONS
211,000

Are penitentiaries a problem? In 1980, close to 500,000 people were in prison and jail.[CH2:N6] By 2017, the number of incarcerated Americans had grown to 2.3 million.[CH2:N7]

The most recent presidential election has uncovered a group of workers who feel they lack the compass to follow the American Dream. They don't feel valued or heard. The tools to attain self-actualization have gone missing. It's as if because the bear has stopped chasing them, they find themselves stuck in their tracks. Millennials are in a different, yet similar, situation. They have the compass, but they've set new terms for the American Dream.

IT'S TIME TO SHIFT THE WORK!

We've won the lottery. We're no longer required to spend all day standing at a conveyor belt. We've achieved a basic level of security. Accessing basic material goods has never been easier. This could empower us to structure workdays that complement our home lives and deepest values.

With all these workers thirsting for change, there is only one thing preventing the next work revolution.

WE DON'T HANDLE CHANGE WELL

Transitions are messy. And not knowing how they will turn out can cause anxiety and fear, and this particular flavor of fear can be overpowering. It can fly in the face of facts and tempt us to stay put and avoid changes. Take people who drive in cars every day, yet are afraid to get on an airplane. Statistics show that the latter is significantly safer, but people are still more fearful of flying because it's a break from their routines.

Fear keeps us stuck in the present, even when the present is no longer stuck. People accept the outmoded idea that satisfaction at work isn't crucial and instead operate with the fear-based model that something terrible will happen to them if they don't take a job that will cover the mortgage of the perfect house in the perfect suburb. This homeostasis is caused by the part of our brain that's designed to keep us in a place of comfortable certainty, even if we're accompanied by misery.

"You can only become truly accomplished at something you love. Don't make money your goal. Instead pursue the things you love doing and then do them so well that people can't take their eyes off of you."
MAYA ANGELOU

BE BRAVE

To make progress, we sometimes need to step into the fire, face our fears, and bravely overcome them.

WHAT'S YOUR SINGLE GREATEST FEAR THAT YOU'VE NEVER SHARED? WRITE IT DOWN.

....

....

....

....

STILL, THERE IS A PART OF US THAT DESIRES SOMETHING DIFFERENT.

Action and horror movies drive the box office because humans possess an innate desire to imagine mysterious, unknown, uncertain, and otherworldly events. Underdog stories like *Rocky* are a favorite because they have unexpected endings. This search for something different is why slow dramas, where not much happens and not much changes, aren't in great demand.

Sameness and difference are engaged in a tug-of-war inside our bodies. People love being married, yet they have difficulty staying faithful. It's like the monkey that doesn't let go of the branch until it has its hand on the next one.

But there's good news: we're not monkeys anymore. The new information we have gathered about the connection between our brains and work can allow us to make educated, science-based decisions about how to shift so we're ALL IN on work and life. It's about placing faith in the facts. Our brains in the head, heart, and gut are also capable of embracing logic and rationale.

The closing lines of Robert Frost's famous poem, "The Road Not Taken," talks about two roads diverging in the woods. Generations of Americans have read these verses as an ode to America, a symbol of its people's courage and independence, our willingness to take the road less traveled, a symbol of the entrepreneurial spirit.

The poetry critic David Orr reads the poem differently, writing an entire book correcting the standard interpretation. He believes the poem is about choice. The reader doesn't know if the narrator's choice of path was arbitrary or meaningful. According to Orr, he says Frost is saying that people will make an arbitrary choice and still tell themselves, and others, that it was a meaningful one and full of courage.

As a society, we are handcuffed by similar dogma. It's a "World Is Flat" moment. We tell myths of our excellence even when the facts don't agree. The United States has the highest incarceration rate in the world. Why is there not more of a national movement to change these numbers, to look at systems in other countries that have proven effective? Solutions are out there, but people are scared that a shift in policy will lead to chaos, even when the data shows it will lead to stability.

IT'S TIME TO PLACE OUR FAITH IN FACTS AND STOP FEEDING OUR BIASES.

Texting and driving is just as dangerous as drinking and driving. A driver who is texting is operating at the same level of impairment as someone who drank four beers before getting behind the wheel.CH2:N8 Yet, despite the evidence, we aren't nearly as outraged over people driving while texting.

In many states, it's still legal! Why? Our love for phones and the emotional comfort they provide closes off our minds to the facts and prevents us from seeing the truth (and tackling this menace).

> *"It is the acceptance of death that has*
> *finally allowed me to choose life."*
> **ELIZABETH LESSER**

DO THE WORK

WHY DO PEOPLE QUIT?

PAUSE HERE; DON'T READ AHEAD.

WRITE DOWN THE FOUR BIGGEST REASONS THAT YOU BELIEVE PEOPLE QUIT THEIR JOB.

Then read on to see how your answers align with supporting data.

For fifteen years now, 70 percent of the workforce has been disengaged and, yet, the business world still clings to the fallacy that money is the chief concern of workers, when survey after survey refutes this claim. Feeling there is no room to grow, poor leadership, a desire for a better work culture, and a wish for more challenging work are the top four reasons people leave their jobs; better pay and recognition at work are reasons five and six.[CH2:N9] Still, ask executives how they seek to get the most out of their workers, and most of them will mention monetary incentives. Money is important, but a work environment that challenges its workers and promotes healthy and congenial relationships between colleagues is crucial. Change at the workplace will not happen until we allow data like this to lead us.

WE SUCK

AT COMMITMENT

Most of us remember the one television set that sat in our family's living room for the entirety of our childhood. My family had one. We also had the same toaster for twenty years, given to us by my grandmother after she used it for twenty years herself.

Nowadays, after owning a television set for more than a couple of years, most people already have their eyes on a newer, flatter, more high-definition model. Nothing anymore is built to last for twenty years because people no longer demand things that will last. We've replaced a generation of fixers with one of replacers. The replacement mentality is seen in our constant technology upgrades, car trade-ins, and pretty much everything else we can get our hands on. We've also lost our sense of curiosity of how things work. Most people don't even know how to change their vehicle's oil. Without the understanding of how something works, we cannot appreciate the miracle of its existence, so we take it for granted.

Our general lack of commitment to objects extends to relationships, too, as seen in divorce rates and employee turnover. As a society, we don't see things through to the end, put in the hard work to make something last, or take the time to figure out how to make an existing product even stronger.

Adding to this problem is our greater access to replacements. We're no longer living on the farm. Our next phone or appliance is a click away. During an average day, we encounter thousands of people who can be our next mate. Like the monkey on the tree, we can hold on to one branch while preparing to grab thousands of others within reach.

We can't even commit to ourselves! Every New Year's, 60 percent of Americans make resolutions surrounding smoking, getting organized, saving money, eating healthy, exercising, and spending time with the family. *Eight percent of people are successful in achieving their goals.* Why can we not make a lasting commitment to treating ourselves better? Why do we spend more time on picking out our clothes and doing our hair than we do on fixing our bodies and souls?

There's a book called *Love Your Life to Death* that I hand out to family, friends, and clients. The author asks her readers to sit in front of the mirror and ask what it is they love about themselves. The point of this exercise is to emphasize that love comes from within ourselves. Is it really our boss's fault that we are miserable? Are we so powerless that we lack the means to overcome what is a preventable situation? Why do we allow other people and objects to define us and determine our level of happiness?

"Perhaps, we should love ourselves so fiercely, that when others see us they know exactly how it should be done."

RUDY FRANCISCO

DO THE WORK

MIRROR, MIRROR

Find a mirror, look yourself in the eyes, and ask yourself this question:

WHAT DO I LOVE ABOUT MYSELF?

Be kind to yourself, and write down the things you said.

Why is it we're comfortable trying to save humanity from starvation and dehydration but squirm when the focus shifts to confidence, self-love, and the ability to self-actualize?

Returning to Richard Saul Wurman's words of wisdom: "Learning is remembering what you're interested in." Commitment only comes when there is passion for the matter at hand. Changing the work world starts with our making the effort to listen to the three brains in our body. Without this, we will have no sense of the world we want to commit to.

WE HAVE
BECOME COMPLACENT

"The opposite of love is not hate, it's indifference.
The opposite of art is not ugliness, it's indifference.
The opposite of faith is not heresy, it's indifference.
And the opposite of life is not death, it's
indifference."
ELIE WIESEL

We despise the job. We despise the work. We don't agree with the company's values.

Yet, we stay.

Don't mistake this inability to move forward for commitment. Loyalty and consistency do not keep us at the jobs we loathe. Complacency is why we stay put, settle, and allow our lives to pass us by, even when we secretly know that more meaningful work is out there.

We can sit at home and blog about the great injustices in the world, and write anonymous posts about our horrible bosses, but how does that create change in an impactful way?

COURAGE MEANS TAKING ACTION, DEMANDING MORE FROM THE PLACES WHERE WE WORK.

If our bosses are acting in a manner inconsistent with the company's mission, we need the courage to say something. If we believe the traditional way of doing business is lacking, we must show a willingness to step outside the lines. Most importantly, we must have the courage to act in the best interests of the group, instead of just looking out for ourselves. The fact that we no longer struggle for the basics can empower us to act for the greater good.

In *The Great Work of Your Life: A Guide for the Journey to Your True Calling*, Stephen Cope writes that the key to finding one's purpose in life can be found in the pages of the two-thousand-year-old Bhagavad Gita. This Hindu text, a dialogue between the Prince Arjuna and his guide Krishna, had a profound influence on historical giants like Harriet Tubman, Henry David Thoreau, Ralph Waldo Emerson, and Gandhi, people who clearly were never complacent and succeeded in finding their purpose in life. According to Cope's reading of the Bhagavad Gita, the journey to discover one's purpose is more important than the destination.

People don't remember the champions hoisting the trophy. They remember the plays that won the game. It's why in movies the hero celebrates only for a minute or two before the credits start rolling. Dharma is this sincere, honest search to discover one's purpose in life. If you pursue teaching, believing this is your passion, and don't succeed, the attempt was not a waste of time or energy. You can still uncover something valuable about yourself in the process, a discovery you can use to continue your search. The key is always to be moving on the path to discovery, and not to allow complacency to keep you stagnant.

LIVE TO WORK

People often claim that work has no connection to their life's purpose. They say things like, "I don't live to work; I work to live." The statement ignores the fact that people spend a majority of their waking day at work. Lottery winners can have the conversation of whether work is a worthy enterprise. The rest of us need to appreciate that even if we put our families first, they still won't be the only important people in our lives. A third of our day is spent with our coworkers. This is far longer than the time we spend with our families.

> *"If you love life, don't waste time, for time is what life is made up of."*
> **BRUCE LEE**

DO THE WORK

DO THE MATH

HOW MANY HOURS DO YOU SPEND ON WORK EVERY WEEK?

WHAT PERCENTAGE OF YOUR WAKING HOURS DOES THIS REPRESENT?

CAN YOU TRULY SAY YOUR WORK DOESN'T NEED TO MATTER?

Also, work and home aren't two separate, unrelated entities. How we fill our time away from the house has a direct consequence on the people we are when we return home. We increase the likelihood of being a good parent if we use our free time to go to the gym, practice yoga, or study mindfulness, as opposed to sitting on in the couch drinking Mountain Dew and eating Doritos. Being healthy doesn't make someone a better person. Rather, self-betterment is a mindset that can positively impact every other area of our lives.

Since there's a connection between the different parts of life—and work plays an outsized role—it's imperative that we find jobs that light us up! Most of us have been around long enough to know that inspiring jobs don't fall into people's laps. It's on us to pursue the ideal job.

> **REPEAT AFTER ME:**
> **"IT'S ON ME TO PURSUE MY IDEAL JOB."**

Money isn't what we're chasing at work. So what is it we're after when we get up and go off to work every day? Dan Pink, in his book *Drive: The Surprising Truth About What Motivates Us*, identifies three elements that must be present for employees to operate at a higher level.

First is autonomy. People see themselves as free and don't want to be ordered around. They won't last in an environment that demands constant conformity.

The second element is mastery. Workers want to feel as if they're advancing and obtaining a level of expertise, as opposed to repeating the same, monotonous tasks day after day.

Purpose is the final ingredient. The job has to mean more than creating a cute widget or racking up billed hours.

3 THINGS THAT MOTIVATE US BASED
ON INTRINSIC MOTIVATION

1 **AUTONOMY**

2 **MASTERY**

3 **PURPOSE**

WHAT DO THESE THREE ELEMENTS HAVE IN COMMON?

They speak to our fear of mortality. Premature death is no longer common. Thankfully, it's the exception. We've survived the evolutionary chain as winners, and we don't take this advancement for granted. Deep down we feel the responsibility to perpetuate life and carry on this project of the human race. We search for ways to extend life even further and increase its quality. It's for this reason that we need to know that our work serves something bigger than ourselves.

WON'T SAVE YOU

For the past fifteen years, businesses have attempted all kinds of management strategies to traverse this evolutionary stage that has workers searching for meaningful work. They thought paying people more money would lead to higher performance and greater loyalty. Some companies believed that lackluster job performance was connected to poor training, so they made retraining a priority. Money and training weren't the antidotes, and now the management consulting space is a forty-billion-dollar industry of which I'm happy to be a member.

Let's not dismiss the slight shift that has occurred in recent years. Many companies have demonstrated openness to wellness and engagement programs, career development, and corporate social responsibility. For many companies, however, these programs and initiatives are nothing but boxes to check off as part of a public relations campaign. They're only flirting with these advancements, unwilling to make a formal commitment to institutional changes. It's one foot in and one foot out, *a sideshow*.

Money is what holds these companies back from building on these changes. They're scared how an environment that fully takes the workers into account will impact the bottom line.

Companies are ignorant when it comes to understanding how people operate. A majority of executives have no awareness of the latest discoveries in neuroscience, which means they aren't using the latest data to inform their hiring and training practices. It's delusional for companies to claim they have a clue on how to screen job candidates. If this were true, they'd have engaged workforces! This cluelessness continues once people are hired, as seen by the way they fumble to devise systems to excite, motivate, and retain workers.

BUT

NEUROSCIENCE CAN

Discoveries in neuroscience can change the way businesses work. For instance, companies no longer need to operate on the outdated notion that people act purely out of rational, self-interest. Researchers studying dopamine—the pleasure chemical of the brain in the head—have observed how it's released when diverse factors like irrational risk taking, altruism, gratitude, and trust are present. The concentration of dopamine in our brain has a direct impact on the choices we make.[CH2:N10]

When we say yes to someone we've done business with before, it's not because it makes sense intellectually. It's actually a release of dopamine from familiarity with the other person that's triggering us to say yes. This understanding of neuroscience has changed how businesses target customers. The next challenge is to use the new research to upend how employers treat their workers.

David Rock, a leadership consultant, coined the phrase "neuroleadership" to describe an emerging field that studies leadership through the lens of neuroscience. It aims to uncover what the science says about self-awareness, insight, decision-making, and influence.[CH2:N11]

This field has discovered a different way to address the issue of engagement. How do we incorporate this newfound knowledge into the workplace if companies are resistant to change, unwilling to face the facts, and unable to look past their own dogma? They believe they're attracting top talent. They're not. They believe morale among the workforce is high. It's not. They believe their employees would accept the job again. The workers would rather work anywhere else.

The few companies aware of the problem don't appreciate its severity, which is why they task human resources with producing a solution. The original purpose of human resources was to oversee recruitment, compensation, and benefits. Sensing a shift in the average worker's needs, HR began handling the training and placement needs of employees as well. Unfortunately, this engagement issue is something HR can't solve, because it's an organizational problem. Accountability for engagement rests mainly with the worker and the worker's direct manager. The people of human resources are not the team leaders in charge of overseeing the work. Asking them to tackle the engagement issue is like blaming the accountants for being over budget. The accountants aren't the ones spending the money, and workers aren't disengaged because of the human resources department.

RESET WORK

The average millennial is content with holding, on average, eleven different jobs before the age of forty-five. Nothing, in other words, will keep them in any position for too long. Being on the move is the objective. So why are companies looking at this desire of millennials to move on as a problem of retention? Why are they coming up with strategies to retain workers forever? Instead, companies can redesign the work environment to conform to this new reality of workers who will stay in a position for only several years.

In the early nineties, Enterprise Rent-A-Car said they wanted to elevate the brand by having college-educated workers sitting behind their rental-car counters. Knowing most people with a degree wouldn't want to spend a career sitting behind a rental-car counter, they recruited college graduates to join a newly established two-year management-training program. After the two years were up, the college-educated managers were free to move on to a different company, and most of them did. These recent graduates signed on for a pretty unglamorous job because Enterprise was offering something valuable: managerial experience.

Businesses know everything there is to know about what customers want through the use of big data, customer relationship managers, and companies like Salesforce.com. So why aren't these same businesses using data to design a workplace their workers want?

Think of stay-at-home mothers. Many of them are well educated and had successful careers before caring for their children. The majority jump at an opportunity to work a reduced schedule. The same is true of seniors. They may be formally retired but have years of wisdom and knowledge to share. Most want to continue working and would make fine part-time mentors. The unwillingness to hire certain kinds of employees because they don't conform to some outdated model of the "nine-to-five worker" reflects a lack of creativity and openness to change.

How many executives stop to consider how many actual hours their full-time employees are working?

DATA SUGGESTS THAT 70 PERCENT OF THE WORKDAY IS UNPRODUCTIVE.

Take away coffee breaks, long lunches, and time on the internet from the employee who is supposedly working a ten-hour day, and the leadership is left with someone who will work far less than the mother who wants to come in, not a take a lunch, and

"They always say time changes things, but you actually have to change them yourself."
ANDY WARHOL

DO THE WORK

SHIFT YOUR SCHEDULE

WHAT WOULD YOUR IDEAL WEEKLY WORK SCHEDULE BE?

Sun.
Mon.
Tues.
Wed.
Thurs.
Fri.
Sat.

WHAT WOULD YOUR TEAM'S IDEAL WEEKLY WORK SCHEDULE BE?

Implement it next week, then return to this exercise and record the following:

WHAT WENT WELL?

WHERE IS THERE OPPORTUNITY FOR IMPROVEMENT?

SHIFT THE WORK

stay off the internet, so she can finish her work before lunch in order to be home in time to meet her children at the bus.

A fellow entrepreneur at a conference I attended, Stephan Aarstol, wrote a book called *The Five Hour Workday*, a plan he enacted for his own company, Tower Paddle Boards. Workers come in at eight in the morning, don't take a lunch break, and leave at one, so they have time to surf. The average workweek is twenty-five hours. What's remarkable is that the daily workload hasn't changed. Workers are not only more productive and efficient with their time, but business is booming, proving that there are other ways to think of the workday and workforce.

Tables are beginning to turn. More and more, workers are succeeding in creating businesses that reflect society. Maternity-leave policy is just one example where employee demands are forcing companies to reform traditional models. Federal law mandates that companies allow new mothers twelve weeks of unpaid maternity leave. Deloitte recently announced that mothers and fathers would receive sixteen weeks of paid leave following the arrival of a baby. Twitter gives mothers twenty weeks of paid maternity leave. Netflix offers new parents twelve months, all paid. Returning parents also have the option to come back as full- or part-time employees.

These new models represent a philosophical shift in which the workers are treated like adults. If a company is going to impose inflexible rules and controls on its employees and not take into account that they have other adult responsibilities in their lives, then leadership shouldn't be surprised when workers act like disgruntled children who are at work under protest.

GREATER MOMENTUM

Because the change is incremental, there isn't enough momentum to shift the level of engagement in a meaningful way. Real change will happen as millennials join the workforce and fill positions where they choose to work like they give a [blank]. Calling them the "me generation" doesn't change the fact that they are the future. The data is clear. They don't care about money, and they have no interest in being lifers at a company. We need to invest new energy into solving this problem, one that looks to neuroscience for answers.

Other thinkers have begun confronting the engagement crisis, including Martin Seligman, the father of positive psychology, who for decades has used the scientific method to explore what makes people happy. The researcher Shawn Achor has written a book titled *The Happiness Advantage: The Seven Principles of Positive Psychology That Fuel Success and Performance at Work*. The last several years have seen the publication of books with titles like *Reinventing Organizations*, *Stealing Fire*, *The Work: My Search for a Life That Matters*, A *Theory of Everything: An Integral Vision for Business, Politics, Science, and Spirituality*, *The Purpose Economy*, *Conscious Capitalism*, and *Evolved Enterprise*.

Clearly, people are taking this problem seriously, and much of their work has influenced my own perspective. This book hopes to add to the conversation by focusing on the science of the neural network connecting the brains of the head, heart, and gut, in trying to craft practical solutions, so workers across the country can answer positively to the question, *"Is what I'm doing worth it?"*

"The ultimate value of life depends upon awareness and the power of contemplation rather than upon mere survival."

ARISTOTLE

PRACTICE AWARENESS

WRITE DOWN THE THREE MOST MEMORABLE MOMENTS IN YOUR LIFE.

What makes each memorable? Notice and record whether a pattern emerges in your answers.

1.

2.

3.

DESCRIBE YOUR IDEAL DAY, WEEK, AND YEAR.

WHAT GETS YOU UP IN THE MORNING?

WHAT KEEPS YOU UP AT NIGHT?

CHAPTER 3

THE FIRE
INSIDE

In the fall of 2000, months after losing my mentor, cousin, and mother, all in the span of less than two months, the doors of SHIFT opened for business. Two weeks later, I nearly died.

Months earlier, over margaritas in El Paso, Texas, a close friend goaded me into competing in a triathlon with him. I'd say yes, I told him, on the condition that he partner with me for this consulting firm we had in mind. We shook hands, and a business was born. Now, all I had to do was train for a triathlon in two months, which would be a challenge, considering I despised running, didn't own a bike, and had last spent time in a pool back in tenth grade when I joined the swim team to impress a girl.

YOUNG, DUMB, AND MOSTLY FULL OF IT, I BARELY TRAINED IN THE FOLLOWING MONTHS BEFORE THE RACE.

On the day of the triathlon, the outer bands of a tropical depression pounded the race area. The forecast promised strong winds and heavy rain as we raced the 1.2 miles in water, 25 miles on bicycle, and 6-mile run to the finish.

A nearly five-foot chop in the water had participants dropping out before the firing of the starting pistol. The organizers managed to place only two out of the ten buoys in the water, meaning it would be difficult to follow the shoreline as we swam two hundred yards out into the water. Officials discussed whether to cancel the event. Unfortunately, for me, they decided it should go ahead as planned.

Loosening up on the beach, I tried to calm my nerves. My father and stepmom, who originally showed up to cheer me on, pleaded with me to drop out. Looking out at the first buoy, I debated the recklessness of swimming the equivalent of two football fields. Once I got to the first buoy, I'd turn left and swim 1,600 more yards. Reaching the second buoy, I'd make a final turn and swim the final 200 yards to the beach. I didn't feel confident about the swimming portion of the race. The little training I'd done was in pools, where I never managed to swim a full mile without stopping to catch my breath. Open-water swimming, I knew, was a completely different beast. On top of that, drowning was a pressing fear, stemming from a scare at age eleven when a lifeguard had to save my life.

BUT A DEAL'S A DEAL, RIGHT?

From the first stroke, waves were battering my body. It felt like I was swimming in place. I put my head down, moved my arms, and kicked my legs at a furious pace. After what felt like an eternity, I reached the first buoy, turned left, and began the long swim along the coastline.

Even after the turn, I kept up my intense speed. Gassed, I finally came up for air. I looked out to the buoy straight ahead to check on my progress. The waves were coming from my right, tossing me up and dropping me back into the water as they passed under me. With water rising all around me, I couldn't locate the buoy. Suddenly, I lost all sense of direction. Keeping my calm, I was determined to find the other swimmers to help reorient myself. Eyes scanning ocean, there was nothing but water in a fifteen-foot radius.

My first thought: nobody was around me because I was winning.

Head back in the water, I started swimming hard. Next thing I know, my hand grazed the ocean floor, meaning I'd moved off the line and swam back to the shore. Now, I had to swim 200 yards out to reinsert myself into the line. When I made it back out, I was out of breath and only halfway done. Intense charley horses stabbed my legs and feet, forcing me to turn over and float on my back.

If we got into trouble, they told us before the race, a hand up in the air would be enough to alert a lifeguard. Swimmers passed me, which felt reassuring, and I knew as soon as I got a second wind, I'd simply follow the other racers. Then, I felt the first sting. Jellyfish, thanks to the storm, swarmed me. They attacked my back. After that they went after my legs. Water lapped over my face. I was panting. The air wasn't making it down into my lungs. Up went my arm.

IT WAS A HUMBLING MOMENT.

I'd never quit anything in my life, and I wanted to cry as I waited for a lifeguard to help me out of the water for the second time in my life.

Soon, it became clear that nobody was coming, and I was struggling to keep my head above water. I was longing to sink. The water felt oddly calm, and I wondered whether the pain of drowning would be felt in my chest or head.

A flood of visions entered my mind. Playing football in high school. Pounding the drums for band. Then, I thought about the moments I'd never experience. Marriage. Holding my own children. My mentor, Stan, and my mother came to mind. Would I be seeing them soon in the next world? Disrupting these thoughts were a hallucination of my dad and stepmom, standing on the shore, grief-stricken as they watched the lifeguards pulling my lifeless body out of the water.

At that moment, the charley horses cleared. My lungs filled with air, my shoulders rising above the water's surface. I flipped over, and my arms and legs began to move.

The decision was no longer about me. Getting back to the shore was something I had to do for my father, who'd spent an entire life making sacrifices for me, his dumb son who had to do a race that he didn't bother training for.

Inch by inch, I proceeded. Out of breath, I rolled over to my back and floated until I was ready to resume. After forty-six minutes, I was the final racer to leave the water.

Three months after the race, I went to swim camp. *A year later, I purposefully entered my second triathlon and cut my finish time in half.*

NEVER QUIT

Being rushed to shock trauma after a football injury to growing up in a neighborhood full of guns and drugs means that I've had my fair share of close calls. The crisis in the water that day was different. I'd given up the fight and decided I was done. Survival became the only option once my body connected with my why.

What was the "why" that filled my lungs with air and got my arms and legs moving? It wasn't those egocentric visions of *me* playing football, of *me* not getting married, of *me* not having children. My "why" was about the *people* who impacted my life. It was only when I saw my father and stepmom, people whose lives I affect, that my body kicked into a higher gear. The vision triggered a physical response. One minute, I'm struggling to keep my head above water, unable to move my arms or legs. Hours later, I'm jogging toward the triathlon finish.

For me, a twenty-three-year-old, the near-death experience put things into perspective about how the most invigorating decisions we make are the ones that take our connections to the wider world into account.

OUR ABILITY TO OVERCOME LIFE'S GREATEST CHALLENGES ISN'T ABOUT PASSION. IT'S ABOUT PURPOSE.

Using our bodies and minds to their fullest potential can happen only when we appreciate that we're something greater than individuals. We pull our greatest powers from our connections and responsibilities to larger groups—family, community, organizations.

SHIFT THE WORK

"The meaning of life is to find your gift.
The purpose of life is to give it away."

PABLO PICASSO

DO THE WORK

STATE YOUR PURPOSE

WHAT IS YOUR "WHY" THAT KEEPS YOU SWIMMING WHEN THE WAVES GET HIGHER AND YOU LOSE SIGHT OF SHORE?

This awareness crystallized when I became a father. Now when I make decisions, I think about having to explain it to my son and daughter in ten years' time. The question is no longer if I'll be satisfied with my choices in the future, but also whether my children will be proud to call me their father.

When I was twenty, I had the opportunity to walk on fire at a Tony Robbins seminar. At the risk of being burned, I stepped onto the hot coals and fortunately walked away unharmed. One reason is because wood is a terrible conductor of heat. There was another reason, though. Tolly Burkan, founder of the Firewalking Institute of Research and Education, says it's more than physics: he credits the state of mind. When a walker is relaxed enough to allow for strong blood flow while continuing to walk, it's as if the mind sparks a fire inside that competes with the fire beneath the feet. So how do you keep the fire burning inside? [CH3:N1]

Adversity will come at you. It's a given. You'll face obstacles and setbacks, professional and personal. If the passion and fire inside stay high, the fires outside can't cause any damage.

DO WITH ENGAGEMENT?

We've made it out of the caves. We're no longer rubbing two sticks together in order to generate fire. We've mastered agriculture and industrial production. We've built cities full of skyscrapers and opportunities for people of all walks of life. Through our achievements we've produced new challenges. Smog fills the skylines of our cities and warms our planet. The rules of basic civility have been thrown out the window, and our politicians can't come together to solve our most pressing issues. Doctors and researchers have succeeded in prolonging life, but this achievement has produced a new concern. The elderly (those who are sixty-five years of age and older) account for a third of all healthcare spending, and the solvency of Medicare and Social Security will be tested in the coming years. CH3:N2

The workforce in America is undergoing major shifts and challenges as well. Baby boomers are retiring, taking with them their honed crafts and institutional knowledge and leaving an enormous talent gap. Meanwhile, inequality in the workforce has become more entrenched. From the years 1978 to 2014, CEO compensation has increased by almost 1,000 percent, while the average worker's pay during this period has risen just 11 percent.CH3:N3 Additionally, white men represent 62.2 percent of private-sector senior-level executives, while black men make up just 1.6 percent of this group. Worse still, while mostly white women comprise a grand total of 24.3 percent of private-sector senior-level leadership, black women only represent a bleak 1.5 percent.

To illustrate, for every 200 people in your company's senior leadership level, you would find 124 white men, 50 white women, three black men, and three black women. According to the statistics, the remaining 20 leaders would be of Hispanic and "other" minority descent.CH3:N4

Assuming we don't have another extinction-level event in the coming years, all these seemingly insurmountable problems will remain, or intensify.

How do we fill the talent gap? How do we get past our subconscious prejudices and allow all types of people to rise to the top of organizations? How do we demand that people stop valuing self-interests over the welfare of our communities?

It starts with understanding the deeper motivation for showing up to work every morning, day in and day out. When I was struggling to stay afloat, my mind went through every reason to live, and it rejected them one by one, until it finally located my higher purpose. When we find that sense of higher purpose that drives us to perform at our maximum level, not only will we carry it with us to work, but we will also take it home and back to our families and neighbors. It will show up in our children's schools, our places of worship, and the ways we give back to our communities. It's the force that will allow us to persevere when the odds seem stacked against us.

"Transformation literally means going beyond your form."
WAYNE DYER

STOKE YOUR FIRE

WHAT LIGHTS YOU UP?

Do twenty-five jumping jacks and then write for five minutes straight! Use your own notebook if you need more room. Go!

WHAT DID YOU FEEL? REFLECT ON YOUR ENTRY.

Americans came together in the days, weeks, and months following 9/11. Firefighters, rescue workers, and medical personnel ran toward the fire to help. Citizens joined the military and opened their wallets. This moment brought out the best in people and underscored our shared humanity. "Never forget," the country declared in one voice. The moment passed, and so did our unity.

How do we remain our best selves without tragedy? What happens when that initial burst of inspiration peters out? Without a way to stay connected to our higher purpose, we will continue to lead lives where every day is a countdown until the end of the workday.

On the other hand, once we connect to our higher purpose, we must work at maintaining it. As workers, no matter our current level of employment, we must keep an open mind to the possibility of reinvention. The rapid rise of technology has already replaced a slew of professions. In some cases, it has altered jobs by removing certain elements of skill. My daughter is seven years old. She will likely never need a driver's license. What does that mean for the millions of truck, cab, and Uber drivers? Computers can review contracts, X-rays, and generate magazine articles. In the coming years, there will no longer be print editions of newspapers. We can debate the time frame, but we can't debate the outcome. If you are under forty and don't know how to code, employment may become an issue for you down the road. Over the last ten years, the world has become a different place, driving forward at breakneck speed with technology at its wheel. The workplace is no exception to these changes. Jobs that require filling now didn't even exist in 2006. If this trend continues, 65 percent of children entering primary school today will ultimately end up in jobs that currently don't exist.[CH3:N5]

In this environment, it's easy for us to lose touch with our higher purpose. The simple response is to turn fatalistic, to think that we and our actions are inconsequential, as if the world is running on automatic and we have no say. The challenge is for us not to let these developments crush our sense of purpose. Starting over stinks, unless we shift our minds and begin viewing these changes as opportunities to go down a different career path. This is our moment to take a role in a nascent industry. This is our moment to start a revolutionary company. This is our moment to play a part in a better future.

> *"The more that you read, the more things you will know. The more that you learn, the more places you'll go."*
>
> **DR. SEUSS**
> *I CAN READ WITH MY EYES SHUT!*

DO THE WORK

DEFUSING THE CONFUSION

WHAT DISRUPTIVE TECHNOLOGY DO YOU KEEP HEARING ABOUT ON THE NEWS, BUT SECRETLY FIND INTIMIDATING OR DON'T UNDERSTAND?

....

....

....

....

Spend ten minutes on Google to get a clearer grasp of what it's all about. Then read on to see how your answers align with supporting data.

Thinkers are already grappling with the question of how we can benefit from these (seemingly) frightening developments. In *Thank You for Being Late*, Thomas Friedman talks about the convergence of disruptions in markets, Mother Nature, and Moore's Law and what it means for jobs. *Bold* by Peter Diamandis explores how disruptive technologies are emerging faster than ever before and how these new companies are looking to impact the world.

The truth is that we're still a long way from fully automated businesses, even if technological advances have come at the expense of many jobs. For the foreseeable future, people will play a part in designing, marketing, and servicing the technology running the world. As the labor market tightens, the disengaged will vote with both their production and their feet. If companies want to continue growing, whether it's a ride-sharing app or a financial firm, they'll need to figure out how to recruit talent from a group that has strict ideals for what they expect out of a job. The heads, hearts, and guts of these new workers will demand that they choose work on purpose.

This isn't theoretical. Some of the largest, most innovative companies are proof of how higher meaning can rewrite the rules of business. Zappos sells shoes. Patagonia sells outdoor clothing. Starbucks sells coffee. The "what" isn't what's exciting about these companies. It's the "how" and "why" that makes them incredible places to work. Outsourcing and automation are not the only stories of how capitalism is being redefined. The social entrepreneurship movement represented by the companies above is real, and it attracts people looking for engaging work.

Life, according to James P. Carse in his book *Finite and Infinite Games*, offers two different types of games. One is a "finite" game, which is structured by "boundaries," or rules. There is a beginning and ending, and the goal of this type of game is to win. In "infinite" games, "horizons" are what control the player. By nature, they are not fixed and always changing. The purpose of infinite games is to continue playing and allow more players to enter the game.

Likewise, engagement is not a finite game. It's not a matter of winning or losing. Instead, it's a game that requires constant renewal. It's a matter of making sure that we are constantly placing ourselves in situations where engagement can happen, as opposed to places that make our bodies sick and make us abandon our higher purpose.

ALL IN

WITH OUR BODIES

Business, for the last 150 years—the way we hire, fire, buy, and sell—has been based on the workings of the brain in the head. We've accepted the idea that people are nothing more than rational actors. This current thinking neglects the ways the brains in the heart and gut shape us as people. We've accepted narrow ways of looking at the body, even when science is painting a different picture.

Ask anyone what makes the human body sturdy, and he or she will tell you it's the bones, muscles, and tissues. Nobody ever answers that it's water, even though water is 70 percent of our body mass. People cannot see water as powerful. They think of the simple combination of two hydrogen atoms and an oxygen atom as a soft and powerless compound. The truth is that, despite its appearance and feel, water is the most powerful compound on the planet; it is lifeblood. It protects life from its inception and sustains it until its last breath. Over time, water can destroy rock. It's truly the earth's final frontier. Seventy percent of the earth is made up of water, and mysteries about the earth's oceans and canyons remain. In those mysteries are possible solutions to the world's great problems.

Similarly, most people think of the brain in the head as what defines us as people. They dismiss the brains in the heart and gut as squishy and soft. This is why when making decisions we are constantly cautioned not to let feelings get in the way. We are directed to make rational choices—use our head brain—as if emotional choices have less value. People struggle to recognize that the gut brain (your intuition) is best used in conjunction with your head brain when making decisions,[CH3:N6] as my crisis in the water taught me.

> *"Everything can be taken from a man but one thing: the last of the human freedoms—to choose one's attitude in any given set of circumstances, to choose one's own way."*
> **VIKTOR E. FRANKL**
> MAN'S SEARCH FOR MEANING

Operating on the basis of only one brain has managed to produce a workforce that is only 30 percent engaged, meaning more than two-thirds of workers find themselves in uninspiring, intolerable work situations.

Based on SHIFT's 2017 All-In Engagement Report, only 15 percent of employees work on what they love to do and what they're good at.

If we got this far by paying deference to only one brain, imagine what we could accomplish when we start acknowledging the power of the other two.

TRUST YOUR INSTINCTS

Think of a recent moment when you used your head brain to make a decision, and ignored your emotions or instincts.

WHAT WAS THE OUTCOME OF THAT DECISION?

WERE YOU SATISFIED WITH THE RESULT?

HOW WOULD HAVE LISTENING TO YOUR EMOTIONS OR INSTINCTS INFLUENCED YOUR DECISION?

SHIFT THE WORK

THREE EQUAL PARTS

In other places in society, we see this focus on the head brain, and the subsequent neglect of the heart and gut brains. For years, we thought IQ was the only indicator of a person's intelligence. Howard Gardner, a developmental psychologist, upended this notion when he introduced the idea of multiple intelligences. The theory is that humans have many ways of processing information, and they aren't related to one another. Someone can have tremendous spatial intelligence and little interpersonal intelligence. A person who has high verbal-linguistic intelligence isn't guaranteed to possess logical-mathematical intelligence. There is, in other words, more than one way to classify a person's intellect.

The Mismeasure of Man, a groundbreaking book by the evolutionary biologist Stephen Jay Gould, also explores the idea that intellect can be calculated by simple measurements like IQ. Standardized tests are meaningless when we consider that not every person taking these tests has the same access to education or is taught the same materials.

Despite the theories of Gould and Gardner, the IQ test is still a widely accepted measurement of intelligence for school selection, job placement, and even capital punishment. Society has failed to accept the other intelligence quotients, like Emotional Quotient (EQ) and Guts Quotient (GQ), which are tied to the heart and gut brains.

The Navy has explored the idea of the three brains carrying equal value, but with most commanding officers being terminated for "bad judgment" and "unprofessional conduct," cultivating leaders' EQs seems to be the answer to retaining top-level officers. Amy Fraher, a retired US Navy commander, believes the solution "for the future of America's armed forces is to foster a new vision of leadership, a perspective less wedded to gender-biased models and more focused on creating an emotionally intelligent warrior."[CH3:N7]

Even the Navy's approach shows limited thinking. They, along with most companies, still haven't accepted the science demonstrating that the three brains make up one complete system. Instead, they are working backward, trying to piece together this neural network.

Widespread change demands approaching human intelligence as Pablo Picasso approached his art.

Before embarking on a work of art, artists traditionally use lithographs to practice the painting or sculpture they hope to create. Each lithograph is an effort to tweak, or fine-tune, a specific element. Pablo Picasso took the opposite approach. In creating The Bull, the master's first lithograph was a polished rendering of a bull. He started, in other words, with the end.

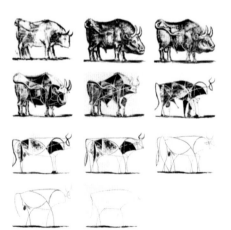

Then, he began to work backward from this fully realistic image he created. He played with the bull's anatomy by adding, subtracting, and changing certain parts. The goal was to strip the bull down to its essence, or spirit.

We too must start from the end—the premise that all three brains play a role in our engagement at work—rather than acting like traditional artists by slowly piecing together a solution. Science proves that all three brains work in tandem.

BRAIN IN THE HEAD

AND IQ

The brain in your heart and gut matter. This fact doesn't diminish the importance of the brain in your head. The head brain has 86 billion neurons,[CH3:N8] the cells that process and transmit information. It's where synapses, electrical impulses, and hormones talk to each other, which is what allows for consciousness and awareness. Most importantly, it's what gives us the ability to identify patterns and make sense of the world.

In business, the head brain is what helps leaders craft strategies to rise to the top. Jack Welch, the former CEO of GE said in his book, *Winning: The Answers: Confronting 74 of the Toughest Questions*, "Winning and losing can't be quantified. They are states of mind, and losing happens only when you give up." This doggedness is the head brain talking. It's what turned Bill Gates's vision in the early 1980s to design the intelligence of all household computers into a reality.

Jeff Bezos is another example. His start was in computers. He identified patterns and trends in the way people were beginning to use the internet. Through his observations, he recognized the potential for an online bookseller. After identifying more patterns about the way people shopped for books online, he saw the possibility of turning Amazon into one of the largest retailers in the world. Despite losing money every quarter, year after year, his head brain provided him with the fortitude to stay confident in the plan he charted.

Amazon and Microsoft have one glaring shortcoming: they don't take into account the brains in the heart and gut. Through the Bill and Melinda Gates Foundation, Bill Gates has helped solve some of the world's outstanding health and education problems. But just imagine how the company may have benefited had Gates decided to incorporate the foundation into the business.

Psychologist and writer Dan Ariely was blasted by a magnesium flare as a young man, resulting in third-degree burns on 70 percent of his body. He spent the next three years restricted to his hospital bed, where he developed a keen interest in people's behavioral patterns. When changing his dressing, the nurse would rip off the bandage in one quick motion, adhering to the belief that it's less painful if done quickly. It sure felt painful to Ariely, and after he was back at MIT, he decided to run an experiment on whether this method was quantitatively less painful. The amount of pain, he determined, doesn't change if the ripping off is done slowly or quickly.

We may think we are acting rationally, when in reality we are listening to the heart brain and gut brain. We rip off the bandage quickly because it's less painful, not for the patient, but for the nurse. In other areas of life, we tell ourselves the best course is the rational one, even though it's influenced more by emotion. How do we fire people? Quickly. How do we break up with people? Quickly. If we were truly using only the head brain, we'd probably come up with a different method.

AND EQ

We've all met people who can describe the inner workings of a jet engine, yet are unable to grasp that they've lost a majority of the crowd midway through their explanation. The head brains of these people operate at a high level, but they have difficulty understanding feelings of their own and others.

Daniel Goleman radically changed the way we think about intelligence, when in the mid-1990s he promoted the idea of emotional intelligence,[CH3:N9] a theory originally developed by the psychologists John Mayer and Peter Salovey twenty years earlier. This theory argues that the ability to know and manage emotions and relationships is integral to achieving success.

Knowing what we know about the heart, it comes as no surprise that it's a center of intelligence. The heart has close to 40 million neurons. It's not nearly as many as the head, but it fills this gap with generating the largest electromagnetic field in the body. When people embrace, their hearts essentially connect, enhancing the transfer of energies.[CH3:N10]

A major function of the heart is to regulate blood pressure, which will rise with passion, excitement, or stress, and may fall with feelings of depression. This is why we feel from the heart. We can consciously make our hearts beat faster by thinking of a decision that excites us. Maybe it's a decision to buy a car we've been wanting, or taking a job that seems bold. We know something in the head, but we own it in the heart. The term "speaking from the heart" implies that what is being said has a certain authenticity, something known on a level deeper than logic. People will actually touch their hearts when they make this statement, as if the knowledge is tangible, and it is.[CH3:N11]

> **IF THE HEAD BRAIN IS ABOUT LOGIC, THEN THE HEART BRAIN IS ABOUT OUR EMOTIONAL CONNECTIONS TO OTHER PEOPLE.**

The heart brain is what defines us as social creatures. *If we are social creatures, what is it our hearts are after?*

The Harvard Grant Study followed the lives of 268 male graduates from 1938 onward.[CH3:N12] At regular intervals, the study, which lasted more than seventy-five years, collected data on their lives. George Vaillant, the study's director, wrote that its key finding is that relationships are the only things that truly matter. A man could have achieved professional success and lived a healthy life, but without love, he wasn't happy. As Vaillant so eloquently wrote, *"Happiness is only the cart; love is the horse."*[CH3:N13]

Most of psychology, through the twentieth century, worked at trying to diagnose what was wrong with people from a psychological standpoint. Martin Seligman, the father of positive psychology, took a different approach. He decided to examine folks who felt fulfilled and content, to learn what works for people. Based on his research, he devised a model called P.E.R.M.A, which is an acronym for the five building blocks of well-being and happiness. The *P* stands for positive emotions, and the *R* is for relationships—being authentically connected to other individuals—meaning the heart brain represents two-fifths of the model.

There are companies that are beginning to understand that good business is about trying to form connections of the heart. Working in a call center, for example, was never thought of as a desirable job. It's tedious. The reason people are calling in is because they're dissatisfied, sometimes angry, and the operator's job is to make the person happy. Spending your days trying to please people is exhausting.

Zappos has taken a different approach to the call center. First, they run the call center out of their offices in Las Vegas, as opposed to outsourcing the work to a different country, like many companies have done in recent years. They wanted the call-center workers to feel part of the company. They also wanted the call center's culture to become a defining trait of the company. Instead of throwing the employees into a cubicle and handing them a phone, all employees undergo four weeks of training. Training is needed because they don't read from scripts when dealing with customers. Operators are asked to access their heart brains, as opposed to their head brains.

All these quirky ways of running the call center come down to helping the workers form a deep personal connection to the company, so they think of themselves as stakeholders and not robots. The call centers are prominently featured in their advertising.

Most call centers practice first-call resolution, meaning they are trying to get callers off the phone as quickly as possible. Zappos's operators, on the other hand, try to keep callers on the line as long as possible. The idea is to build relationships, not make sales. There are some wild stories proving this point. One customer service rep traveled to a rival store in order to purchase for a customer a pair of shoes that were out of stock on the Zappos website. Another operator overnighted a free pair of shoes to a groom who had forgotten his pair back home.

The record for the longest call at the Zappos call center is 10 hours, 43 minutes.[CH3:N14] Tony Hsieh, the founder of Zappos,

"I know it seems hard sometimes but remember one thing. Through every dark night, there's a bright day after that. So no matter how hard it gets, stick your chest out, keep ya head up...and handle it."

TUPAC SHAKUR

DO THE WORK

WRATH TO REWARD

WHICH PART OF SERVING YOUR CUSTOMERS DRIVES YOU NUTS?

WHAT WOULD IT TAKE FOR THAT TO BECOME THE MOST REWARDING PART OF SERVING THEM INSTEAD? WRITE IT DOWN.

SHIFT THE WORK

understands that customers call these centers, not because they are angry, but rather they're lonely and want to talk to live people. As one of the senior brand-marketing managers said about the company, "Our biggest efforts revolve around building likeability around our brand so that consumers turn to a brand that they trust, find reliable, and have an *emotional connection* with."[CH3:N15] It's a strategy that's all about the heart brain.

As much as we love automation and technology, we love connecting with people even more. It's the brilliance of Facebook, a platform that allows people to exchange energy with like-minded people.

The mainstreaming of the mindfulness movement has millions of Americans finally taking the risk to confront emotions and feelings instead of running away from them or locking them away. More people are willing to discuss the complexities of relationships with loved ones and colleagues. Society has begun to view sensitivity as strength not weakness.

As Brené Brown says, "Embracing our vulnerabilities is risky but not nearly as dangerous as giving up on love and belonging and joy—the experiences that make us the most vulnerable. Only when we are brave enough to explore the darkness will we discover the infinite power of our light."[CH3:N16]

THE GUT

AND GQ

We may speak from the heart, but we feel it in the gut. This brain consists of two nerve centers called the myenteric and the submucosal, which have approximately 100 million neurons. This is more than the spinal cord.

The gut produces 70 percent of the hormone cortisol, which is released during stressful periods. It does more than turn our stomachs at anxious moments. Cortisol regulates metabolism, controls blood pressure, and assists with memory formulation. The gut brain is also responsible for processing information during sleep. Seventy percent of serotonin—the neurotransmitter responsible for relaying signals across the brain to help us think clearer—is produced in the gut. The gut is also responsible for 90 percent of our immune system. Your brain in your gut keeps you healthy!

When we talk about finding your greater purpose, we are talking about this intelligence center. When we are faced with a crucial decision, and we appeal to our intuition, we feel a tightening in our stomachs. Those butterflies-in-your-stomach feelings before giving a big talk, sitting down to an interview for a dream job, or walking down the marriage aisle is the gut brain's way of talking to us and telling us what's important.

It's only in these deepest parts that we know our greater purpose, and yet we've been taught to think that it's nonessential, or something secondary, to the tasks of living and working—similar to how companies view their charitable efforts. They see them as something outside the actual business, or, if we want to be cynical, part of a public relations effort.

Nonetheless, there are companies like TOMS Shoes that are willing to put their money where their mouth is. TOMS Shoes entered the highly competitive foot-apparel market with a mediocre shoe. Nobody who buys a pair of TOMS Shoes is going to rave about the shoes' quality. Customers buy a pair because they want to be part of a cause. The company's "one for one" movement donates a pair of shoes every time a purchase is made. Since the start of the program, TOMS Shoes has donated more than 50 million pairs of new shoes

"The Master said, 'If your conduct is determined solely by considerations of profit you will arouse great resentment.'"
CONFUCIUS

DO THE WORK

PURPOSE AND PROFIT

WHAT IS THE CURRENT PRICE OF YOUR FLAGSHIP PRODUCT?

WHAT WOULD THE NEW PRICE BE IF IT WAS 25 PERCENT HIGHER?

WHO WOULD/COULD YOU HELP WITH SOME OF THAT EXTRA MONEY?

to disadvantaged children. The same is true for their other products, and the company provides water, sight, and safe birth to people in need.

Millennial Marketing tells us that almost 50 percent of all millennials would increase their enthusiasm for spending if their purchase supports a cause. They want buying to feel like giving. CH3:N17

CLIF Bar & Company's "Five Aspirations"—sustaining our business, sustaining our brand, sustaining our people, sustaining our communities, and sustaining our planet— guide the company's decision-making process. Customers understand that they are buying far more than an energy bar with each purchase.

Harley Davidson struggled mightily back in the early 1980s when Japan's manufacturing took the upper hand. After several management buyouts, the company was down to a 15 percent share of the marketplace. A new CEO took over, and the first thing he did was leave his office for an extended period. He traveled the country, visiting dealerships and hosting Harley-sponsored road rallies, which brought thousands of bikers together for the purpose of celebrating Harley Davidson. This new strategy was about reconnecting customers and dealers to this product. His idea was that Harley riders thought of the motorcycle as more than a bike. They thought of it as a community, one that is based on principles of patriotism and the independent spirit. Can we name another company whose name is tattooed on more people's bodies? In less than fifteen years, market share has climbed to 85 percent.

Angela Lee Duckworth taught math to seventh graders in a New York public school. Several months into the school year, she noticed that some of her strongest performers had low IQ scores, while some of her smartest students weren't doing well. After graduating with a master of science in neuroscience, she pursued a PhD in psychology, focusing on why some children and adults succeed and others don't.

She and her research team went to West Point Military Academy to predict which cadets would stay in military training, and which ones would drop out. They went to the National Spelling Bee and tried to forecast the students who would advance furthest. They partnered with private companies and tried to see which salespeople would earn the most money for the company. In the end, one trait surfaced as a predictor of success in all these cases: *grit.*

Duckworth defines grit as, "passion and perseverance for very long-term goals. Grit is having stamina. Grit is sticking with your future, day-in, day-out, not just for the week, not just for the month, but for years, and working really hard to make that future a reality. Grit is living life like it's a marathon, not a sprint."[CH3:N18]

When we are failing and feeling our weakest, the only way out is to reconnect with our larger purpose, or our gut brain.

"Did I live? Did I love? Did I matter?" These are three questions we will face at the end of our lives, according to Brendon Burchard, the founder of Experts Academy and survivor of a near-death experience. Gut-level exploration at its finest.[CH3:N19]

WHAT DOES SUCCESSFUL CHANGE REQUIRE?

A BETTER YOU!

THE HEAD BRAIN IS ABOUT INNOVATION AND EXPLORATION.
Do you see a path on the road less traveled?

THE HEART BRAIN IS ABOUT INSPIRATION AND OWNERSHIP.
Can you maintain enthusiasm and excitement as you travel down that path?

THE GUT BRAIN IS ABOUT IMPACT AND THE GREATER GOOD.
Why are you going down this path anyway?

SHIFT THE WORK

CHAPTER 4

TIMELESSNESS

WILL SET YOU FREE

PUT YOUR CALENDAR WHERE YOUR MOUTH IS. IN OTHER WORDS, YOUR IDEALS AREN'T WORTH A DAMN UNLESS YOU HAVE A CONCRETE ACTION PLAN.

SHIFT THE WORK

HOW THE BRAIN IN MY HEAD MADE

GROW REGARDLESS
A BEST SELLER

I was intent on writing a book. Here were the problems: I'd never written one, I didn't have a platform, and more than fifty agents and publishers told me they weren't interested in my proposal.

Yet I was able to overcome these obstacles, out-innovate the industry, and get my first book, *Grow Regardless*, on the *New York Times* best-seller list.

Writing the book—believe it or not—was the simplest part. It took only ninety days (this book, on the other hand, took me eighteen months).

Then came the hard part. For the next year, our team studied how to successfully launch a book with no industry backing. Step one was to learn about the world of books. Weeks into our investigation, we uncovered a harsh reality: if a book doesn't reach best-seller status straight out of the gate, chances are it will die on the vine.

Next, we discovered more depressing facts. Annually, more than 300,000 books are published through traditional channels. Adding self-published books, the number of new titles each year approaches a million, which comes out to around 83,000 books a month. Meanwhile, sales are stagnant. The average nonfiction book in the United States will sell fewer than two thousand copies in its lifetime.[CH4:N1] If only 20 percent of people who buy the book end up reading it, an author is lucky to find four hundred readers—a tough realization when considering how I invested blood, sweat, and tears into writing the book.

It didn't take long for the team to stumble on a glimmer of hope.

The industry, we soon discovered, grades on a curve. See, a book doesn't need to sell a ton of books to make best-seller lists. It only needs to outsell the other books on the market at the time. Normally, this is still a herculean challenge since major publishers can spend up to several hundred thousand dollars promoting their books, doing everything possible to gain maximum exposure, like booking the authors on morning talk shows and buying ad space in major publications.

Knowing we couldn't compete with the major publishing houses, we put our head brains together and designed an innovative approach. We aimed for minimal exposure. This would become our advantage. In selecting a release date, we chose the shortest month of the year, in the worst week of the worst quarter for book sales. We released *Grow Regardless* the morning after the Super Bowl.

In selecting the coldest month of the year, in terms of books sales and weather, we'd avoid competing against the year's hottest books. Instead, we'd make the best-seller lists by standing out and outselling the weakest books on the market. The lists would then become our validation and greatest marketing tool.

Before the release date, we executed a strategic presales campaign, getting the book into the hands of people who mattered. A little luck also never hurt anyone. At a moment when the federal government was passing the unorthodox sequestration, the subtitle of our book, *Of Your Business' Size, Your Industry or the Economy...and Despite the Government!*, surely tickled people's imaginations. The stars aligned, and we hit all of our goals within the first month. Then, we hit every major best-seller list.

Writing this book was not about making money (proceeds were given away to charity). Our aim in writing the book was to get one hundred thousand people back to work. It was a David-versus-Goliath story about how some of my greatest life and business mistakes could teach us lessons about changing the cultures of organizations. The book was written from the heart and gut, yet it was the head brain that allowed us to outsmart the industry and actualize a movement by getting the book in people's hands.

Looking back, my head brain was working on this book before I even opened the doors to my company. At twenty-one-years old, I'd sit and draw images of the future I imagined. In the drawings, I was in good health and ran multiple businesses and a foundation. In several of the drawings, I was writing a book.

Success isn't built overnight. It's not even built after you've spent three months writing a book.

THINKING FAST VS. THINKING SLOW.

The head brain has two modes of thought, according to economist Daniel Kahneman. "System 1" is thinking fast. Something happens, and we feel the need to respond right away without considering consequences. This type of response is based on emotion and instinct. "System 2" is slow thinking. This is when we take time to react. The reaction is deliberate and is based on logic.

If our team had operated under the fast thinking of "System 1," we probably would have released the book right as it rolled off the press. Instead, we were deliberate, carefully weighing the data and using it to drive our decisions. Not an easy feat, considering I'm a pretty impatient guy. Here, I had this tremendous accomplishment in finishing a book, and I had to keep it in my back pocket for an entire year.

Success is the product of a well-thought-out plan. It's also about embracing timelessness, as opposed to being a slave to the clock. Innovation happened because we didn't allow ourselves to become tied to an arbitrary calendar. This state comes from the head brain. If you want to succeed, you need to be deliberate about both the journey and the destination.

THE BRAIN IN THE HEAD

SETS US ON A PATH

Let's face it. We try to tell ourselves that success is subjective, but we live in a world where success matters a great deal. People started looking at me and the company differently after the book became a *New York Times* best seller. Soon, the success threatened to change the dynamics of the company.

The book was a team effort. A writer and several editors helped me with the writing. My entire company put in years of work with our clients, which was the basis for the entire book. Yet, Fox Business and Bloomberg were interested in interviewing only me. I was the one invited to speak at conferences. Professionally, it put me in a different light. Fame had its way with my ego, even if I was constantly questioning whether I'd earned it. Under these circumstances, it was easy to lose perspective. The stomach infection that developed in the second half of 2013 is evidence of how the success changed me both physically and mentally.

The stomach infection was a gut brain problem, but any solution, or cure, demanded I reconnect with the head brain, which had successfully led me to this major milestone, which I slowly began to understand was merely an early stop on a long journey. Without the head brain, our team never would've come up with a successful game plan to survive in the win-lose environment of publishing. Publishing is a win-lose environment, survival of the fittest, and the head brain's ability to identify key patterns and data allowed us to come out on top. It would be the head brain that would ultimately learn how to own the success and use it for a greater purpose.

NEUROSCIENCE TELLS US ABOUT ENGAGEMENT

The head brain itself can be further deconstructed into three separate brains.

In the center of the brain we find mini-brain number one, also known as the reptilian brain. It handles the ultimate task of staying alive by regulating temperature control, hunger, and our fight-or-flight response. It's the part of the brain that keeps us safe from predators.

Wrapped around the reptilian brain is mini-brain number two, or the limbic system. It's more complex than brain one. It's responsible for our emotions, and we share it with other mammals like cats and dogs.

Humans are one of the few species who have mini-brain number three, the frontal cortex. Wrapped around brain number two, the frontal cortex allows higher-level thinking, like engaging in complex social interactions and making plans for far into the future. The frontal cortex is what enables us to innovate and act with intention.

Breaking down these three parts of the brain, we see a direct correlation to Maslow's hierarchy of needs. The reptilian brain is suited to secure the bottom two levels of the hierarchy, which are basic needs like security, food, and warmth. Climbing up the hierarchy, our needs grow to include psychological ones, like forging relationships, feeling accomplished, and being respected. The limbic system, and its management of feelings, is key to obtaining these mid-level needs.

SHIFT THE WORK

The frontal cortex, the thinking part of our brain, is responsible for realizing the uppermost level of Maslow's hierarchy of needs, which is self-actualization, the idea that we can achieve our full potential. As long as our progress isn't stopped by an inability to meet the needs of the bottom of the pyramid, we are free to create a life that is directed at making our ideals a reality.

MODERN WORKPLACE

Most modern work environments are still based on models dating back to the industrial revolution. Don't let the Ping Pong tables and stocked break rooms fool you. Conformity is still king in most organizations. The employer-worker relationship is designed to satisfy our reptilian brains. Do the work and get paid, so you can pay your rent and put food on the table. In such an environment, workers aren't encouraged to think for themselves, let alone pay attention to their emotional needs or desire for self-actualization. Needless to say, this isn't a recipe for empowering employees. It's difficult for a disempowered worker to feel engaged.

Nobody has informed these companies that most workers are no longer in constant fight-or-flight mode, worried that a change of jobs will leave them hungry and on the street. We've ceased worrying about the bear eating us, which frees us up to focus on tasks bigger than ourselves.

In the workplace, the common belief that humans use only 10 percent of their brains may be accurate. That's unfortunate, considering our brains have the capability to operate like supercomputers, processing more than seventy thousand thoughts in the course of a day. Companies are basically telling workers that a majority of the information produced is useless, and may even come at the expense of a job well done. They are, in effect, encouraging workers to function at a fraction of their brain power. Without an outlet for these tens of thousands of thoughts, it's no wonder that we're diagnosing a record amount of people with ADHD. Seventy to 80 percent of what happens in our brain is based on the things we see. Unless we put blinders on people, we're never going to stop the flow of thoughts.

There are real myths surrounding the power of the brain, and the reality of the brain's power demonstrates the need to allow it to reach its potential at work. People's brains don't start deteriorating after the age of forty. In certain areas—linguistic skills—the mind actually improves with age. Also, we're not hardwired. In fact, the brain is amazingly plastic, able to learn new skills, create new pathways, and rewire itself effortlessly. CH4:N2

What if, instead of medicating people, we figured out how to strengthen this muscle in the head so it could be incorporated into the workplace? Used wisely, the frontal cortex is a tool to drive innovation and find solutions to our greatest problems! It's one-third of the brain puzzle that will help us shift the work.

"Self is a sea boundless and measureless."

KAHLIL GIBRAN
THE PROPHET

DO THE WORK

LIMITLESS LEARNING

What is something that struck you as fascinating in your workplace? Read one article about this topic today.

JOT DOWN WHAT YOU LEARNED.

....

....

....

....

....

....

....

....

....

....

....

....

Do this every day!

HEAD BRAIN FREE

Innovation answers the question of how to get from point A to point B in a more efficient manner. It happens after a worker spends time in the saddle, devoting energy and attention to a specific task. Only then can the worker recognize how it can be done in a superior way. The reptilian brain, by design, isn't interested in innovation. It will always take the road more traveled because its sole interest is in guaranteed results. Obviously, a workplace can't entertain every idea that passes through a worker's head, but the solution isn't to completely shut down the person's mind. It's a question of using the frontal cortex—this supercomputer—to the company's advantage, so the worker can become empowered and stay engaged.

My dad worked at the same company for forty years. He didn't have time to innovate. He had bills to pay. Not using his frontal cortex on a day-to-day basis didn't turn him into a dissatisfied worker, because he didn't know better. Studies show that despite their meager lifestyles, indigenous populations experience high levels of happiness. Without access to the outside world, they don't know what they're missing.

THAT WAS THEN, THIS IS NOW, AND NOW WE KNOW BETTER.

We've witnessed the possibility of living a different kind of life, and yet we've found a way to be miserable for this awareness. We pretend we're stuck on the bottom rungs of Maslow's hierarchy. We tell ourselves that we're barely surviving, skating by every month, all because we're afraid of embracing the possibility of becoming innovators at work. The thought of changing jobs, careers, or mindset in a current role seems terrifying to most of us.

Putting food on the table thirty years ago was the primary challenge for many American families. In this day and age, the struggle boils down to finding coveted varieties of coffee that will impress our friends. Owning a boat used to be a big deal. Now it's about owning an island. The first step to becoming innovators at work, and in life, involves stepping outside this posture of poverty through an honest reassessment of our material lives. What if we gave up our need for brand-new cars, premium cable, the latest iPhone, and a massive house? Once we narrow our definition of needs, we can finally turn off our reptilian brains and turn on our frontal cortices.

> "The Master said, 'If your conduct is determined solely by considerations of profit you will arouse great resentment.'"
> CONFUCIUS

DO THE WORK

NIP SOME NEEDS

WRITE DOWN FIVE THINGS YOU'D LOVE TO BUY BUT HONESTLY DON'T NEED RIGHT NOW.

1. ....

2. ....

3. ....

4. ....

5. ....

ECONOMY

Recently, I had the privilege of hearing Gary Vaynerchuk speak at an exclusive conference in Ojai, California. The message he delivered that day to 130 entrepreneurs from around the world was that the world has advanced. Businesses no longer corner the market through location or exclusivity of information and data. We've moved past the information age and creative economy. Now, success depends on one's ability to effectively engage customers and workers.

This is how, in the early days, he transformed his father's liquor store into an online venture that did tens of millions of dollars in sales five years after the site started. It's not that he simply figured out a way to sell wine online. Rather, he developed a brand. He did blogs, which were revolutionary at the time. He also created Wine Library TV, one of the first video blogs to hit the internet. This platform gained him a huge following, partly because of his knowledge but also because his personality connected with people.

In understanding that his success was innovating ways to engage customers, it makes sense that his next venture wasn't another retail internet business. Instead, he started a company that teaches Forbes 500 companies—old, calcified institutions— how to incorporate social media strategies into their businesses, so they too can come up with innovative strategies to engage customers and workers.

Vaynerchuk was receptive to changes in the marketplace. Instead of denying the data, and staying stuck in the past because of an unfounded fear of survival, he took advantage of the opportunity. Like Bill Gates, Jack Welch, and Jeff Bezos,

SHIFT THE WORK

Vaynerchuk fully utilizes the power of his head brain, particularly the frontal cortex. He's managed to free all of his minds from the constraints of patterns and time to embrace what it means to be human.

> "How don't you stay driven... You won the lotto of the universe. You are a human being on earth... You're not some weird specimen in Mars that does nothing. You're not a bird. You're not a sunflower... You won the ultimate lottery, and you're not driven, and you're not fired up, and you're not going for it, and you're willing to sit like a lump and just wait till you die? You suck."[CH4:N3]
>
> **GARY VAYNERCHUK**

He's right. We should consider ourselves lucky. The right sperm met the right egg in order to create you. The probability of you being born was 1 in 102,685,000.[CH4:N4] Many people and companies are beginning to appreciate their uniqueness. They welcome the miracle of life, and refuse to throw it away! It's the immigrant mindset of making sure that every waking moment counts, whether it's at work, at home, or at the store buying dinner.

Building a website and paying for a few Google ads doesn't cut it anymore for businesses. The success of a company like Whole Foods is based on their ability to tell a story about the food on their shelves. It's not just an apple, but an organic apple picked from a family farm upstate. The innovators at these companies are utilizing their head brains in order to connect to the head, heart, and gut brains of consumers.

HOW WE USE OUR BRAINS AT WORK

CLARITY OF PRIORITIES. The brain in the head is constantly assessing how to address the stream of new ideas, and competing tasks we believe to be our responsibility. Our once-extensive attention span has dropped from 20 minutes to a mere nine seconds. The world is changing faster than we can keep up with, and it's influencing our thinking. The truth is, to make and manage our to-do lists, we must know and understand our priorities, the things that are truly important for us to take on, including in what order and in what measure. The sorting responsibility isn't ours alone; managers and companies also must be clear on the priorities. Agreement on and space to focus on those priorities contribute to maximizing engagement. But, according to SHIFT's 2017 All-In Engagement Report, only 23 percent of employees are clear on the priorities that drive high performance.

ONLY 23% OF EMPLOYEES

ARE CLEAR ON THE PRIORITIES THAT DRIVE HIGH PERFORMANCE.

We have to move past our existing paradigms and belief systems in order to embrace work with a fresh perspective. This means being open to learning new things that cater to our curiosity, inform our priorities, and drive our performance. When we learn something, our brain seeks to repeat the pattern over and over with hopes of achieving the same result. This is repetitive learning. To develop the neuroplasticity of the brain, we need to create new bindings. We often hear how older people should do crossword puzzles to keep the mind active. It's a good memory exercise, but it's not a good thinking exercise. Learning how to play chess, or trying to achieve higher levels of skill with an instrument, is what creates neuroplasticity. They say you can't teach an old dog new tricks, but this isn't true. We live in a world where many of us stop learning after we leave college. We are in jobs where we're set on automatic. There's reason to think this stagnation is a contributor to the plaque between synapses, which is the cause of Alzheimer's. Once we appreciate the brain in the head's ability to learn new things, we can welcome the possibilities to learn and grow.

COMMITMENT TO PASSION. The brain in the heart instigates our struggle with commitment. Leading with fear, as opposed to love, has caused our heart to give up rather than fight back or seek resolve. This is evident in divorce rates (53 percent in the United States), employee turnover, rapid technology upgrades, and frequent car trade-ins. As a society, we need to reconnect with what we believe is worth fighting for and then put in the work of fighting for it. SHIFT's own research on engagement has revealed that less than half of the population—only 30 percent of employees—is passionate about the work they do.

ONLY 30% OF EMPLOYEES

ARE PASSIONATE ABOUT THE WORK THEY DO

We must own and feel the passion we have for our work and allow the intensity of our feelings help us achieve our top priorities.

EXPERIENCING PURPOSE. The brain in the gut is torn between trust and distrust, influencing our decisions and reactions. It's what triggers our protective fight or flight instinct. Yet, with a 70 percent disengaged workforce, of which 50 percent are sleepwalking through their jobs and 20 percent are downright miserable, this tells us that those people (and their employers) are ignoring basic instincts. We stay in meaningless jobs, mistaking complacency for commitment, having lost sight of our purpose in life. To be completely engrossed in what we do, we must know that we are having impact and be able to drive our purpose inside and outside of the workplace. The importance of purpose cannot be overstated. It grounds us, provides meaning, and serves as a guide to establish priorities, achieve goals, and go ALL IN. Yet, SHIFT's engagement data shows that only 24 percent of employees believe their mission and purpose is compatible with their company's. We spend one-third of our lifetime at work; shouldn't it light us up?

ONLY 24% OF EMPLOYEES

BELIEVE THEIR MISSION AND PURPOSE IS COMPATIBLE WITH THEIR COMPANY'S

"One love, one heart, one destiny."
BOB MARLEY

DO THE WORK

WHAT'S YOUR DESTINY?

Write down the five things that...

YOU LOVE TO DO:

1. ....

2. ....

3. ....

4. ....

5. ....

YOU'RE GOOD AT:

1. ....

2. ....

3. ....

4. ....

5. ....

YOU'RE PAID TO DO:

1.
2.
3.
4.
5.

THE WORLD NEEDS:

1.
2.
3.
4.
5.

WHAT OVERLAP DO YOU NOTICE ACROSS ALL FOUR CATEGORIES?

This is your purpose! Make an effort to incorporate this into your work each and every day.

CHOOSE YOUR REALITY— BETTER YOU.

Survey the small percentage of people who succeed at following through on their New Year's resolutions, and you'll find that most didn't rely on tangible tools to reach their goals. They didn't use the patch to quit smoking or surgery to lose weight. Instead, they used their frontal cortex to change their perspective. They dealt with the issue head-on, thinking about why they overate and all the moments in life that challenged their self-control. This allowed them to come up with an effective strategy to avoid temptation.

IN SHORT, PEOPLE CHANGE THEIR LIVES WHEN THEY MOVE AWAY FROM A MODEL WHERE THE PAST DICTATES THE FUTURE.

It's not easy to put one's mind in the present. It takes enormous effort to attain the mindfulness to stop worrying about the future or regretting the past. A shift in how we position our mindset, so it's with us in the present, starts with how we treat our mornings.

A regular gym, yoga, or meditation routine is an excellent way to slow down the pace of our thoughts and move into this new mindset. It's preferable that we take care of it first thing in the morning. I like to think of my days as airline flights, and I'm the captain of the plane. Most of us wake up in the morning and take off as quickly as possible. We shower and dress in record time, and grab a bagel with cream cheese to eat in the car, or we just wait to buy a Danish and coffee from a place near work. This is the equivalent of taking off and waiting until you're in the air to think about a flight plan or destination. The rest of the

MINDSET MODEL

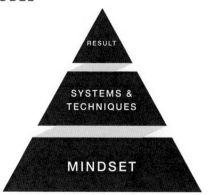

Back to the mantra: what you focus on, you will find.

day ends up being a game of catch-up. Forget preparing for a meeting. We're lucky if we manage to show up on time.

As the pilot of my airplane, I'm a stickler for routine and process. I want to examine every instrument before takeoff, to make sure it's in good working order. I certainly want to identify the destination before the plane speeds down the runway. To achieve this aim, every morning begins with yoga, the gym, or meditation. As someone who runs a business and has two small children at home, I can tell you that committing to a routine is painful at times. What pushes me forward is anticipating the feeling I'll have at the end of the day, when I see that I've accomplished what I set out to do that morning.

It's not only about starting the day right. Throughout the day, I'm constantly doing everything possible to slow down time and stay in the moment. It's a process similar to the one outlined by Atul Gawande in his classic *The Checklist Manifesto*, a book that details how checklists are an effective way to save lives. If I have a meeting later in the day with ten people, I'll set aside one hour to prepare. If I'm going to deliver a keynote address in front of a hundred people, then I will devote anywhere from five to ten hours on my remarks. Preparation time should always exceed the length of the meeting. I'm using this time to organize my thoughts and put things down in writing. This is more thorough than trying to organize your thoughts as you run down the hallway on your way to the meeting. There's far less chance of having anxiety before the meeting starts or regrets once it's over. Instead of scraping through the day and hoping instinct will carry me to success, I'm utilizing the amazing section of my brain called the frontal cortex to my advantage.

Years ago, our team was sitting with a client, the chief risk officer at a bank. He began grumbling about how much he dreads the employee reviews. He called it the single worst part of his job, even though he did it only once a year. At some point, I finally cut off his complaining.

"Look," I said to him. "You don't *have* to do this. You make more than seven figures a year. You can go get a different job. But if you plan on staying in this position, then honor this responsibility. Step outside yourself for two seconds, and consider what these reviews mean to the people you lead on a daily basis. This will be the only time during the year that they will hear concrete feedback regarding their job performance. Your words will have a major impact on the course of their careers take moving forward. They'll be nervous in the days leading up to the review. They'll have people in their lives— spouses, parents—that will be anxious for them. They'll take time preparing arguments and viewpoints. You don't swing an ax for a living, or spend your days sweating under the sun. You work inside a bank because you had the good fortune of being born to a good family and going to excellent schools. You don't have to do this. You *get* to do this."

He mumbled and grumbled under his breath, uninspired by his great opportunity. Eventually, he was fired.

Every moment is an opportunity to execute your vision. Don't waste it on what came before, or let it pass you by because you're too fixated on what comes after.

CONSCIOUS EFFORT

Workers are like drivers heading down a dark road, and the company's values and mission are the headlights that show us a path forward. Even if the headlights are only strong enough to show us what's fifteen feet ahead, we still feel confident continuing down the road because we have faith that the company's mission will keep guiding us through the darkness.

To have any chance at engaging the head brain, we need to work for organizations that have clear visions, and our roles in contributing to the mission should be clear. We must feel that we are provided with the space to use our frontal cortices to innovate in our jobs. It's about the things that wake us up in the morning, and the ones that won't let us sleep at night.

Core 4, a leadership exercise designed by the consultant Gerry Sandusky, starts with a few simple questions: When you meet someone, what is the impression you want to leave on him or her? How do you want other people to see your vision and values? What's your essence?

Several of the major technology companies that have launched in the last generation are steadfast when it comes to the question of mission. Google, for example, has bought tons of other companies, some of them major, like YouTube. But they haven't gotten away from their original mission. Even with their many major acquisitions over the years, the look of the web page hasn't changed.

Taking the issue of mission and values a step further, here are two more companies whose missions and values act as guideposts for how workers move through their workdays.

Glassman Wealth Advisors (a great client of ours) devotes one day per month for employees to sit and think. Football teams hire strength and conditioning coaches and set aside time for players to hit the gym. Yet, how many companies dedicate time, space, and personnel for employees to work on strengthening their brains?

"Walk as if you are kissing the Earth with your feet."
THICH NHAT HANH
PEACE IS EVERY STEP: THE PATH OF MINDFULNESS IN EVERYDAY LIFE

DO THE WORK

THINK ON PURPOSE

Schedule one hour in your day tomorrow to sit comfortably and think. No smartphone. No laptop. No conversations. Just quietly spend time with your own thoughts.

Zingerman's Deli in Ann Arbor, Michigan, has what they call their "Three Bottom Lines." Bottom Line #1 is Great Food. Bottom Line #2 is Great Service. Where it gets innovative is Bottom Line #3: Great Finance. Every worker is supplied with detailed information about the company's finances. This effort at educating the workforce is the very definition of running a culture that treats everyone like adults and partners in the mission. Employees who know the profits and losses connected to their specific job are more likely to innovate and produce creative ways to do the job in a more efficient and productive manner. In this environment, the person who cleans the trays off the table is more likely to bring up noticing how customers throw out a lot of unused napkins and ketchup packets.

RICH PEOPLE HAVE MONEY,

WEALTHY PEOPLE HAVE TIME

In most professional services that bill clients by the hour, employees are expected to account for more than 80 percent of their time at the office. If someone is spending forty-five hours a week at work, then, at the minimum, thirty-six of those hours should be spent working. Workers cannot keep up this pace for long without burning out. Keeping up with this breakneck tempo certainly leaves little time or energy to innovate. If we spend all our time working for others, when and where will we work on ourselves? Not at home after working a nine-hour day (not including the two-hour commute).

In 2009, at the height of the recession, our company hit a wall. With the slowdown in business came the harsh realization that we weren't as good as we thought we were. We had to figure out how we were going to dig ourselves out of the hole.

How do people get better? How do people become the best? Last time I checked, they practice. Some companies have mandatory training once a year. Other companies hold seminars once a quarter. We decided to devote time to focusing on ourselves. Every Friday we reserve time as a team to focus on our whole self and to engage in a company-wide activity that changes from week to week. We've done massage therapy, acupuncture, yoga, and meditation. Some weeks we will serve the community, handing out meals to Baltimore's homeless or planting community gardens. We've invited speakers of all kinds. Occasionally, we'll gather the entire company and present a current project that is proving difficult for one of our teams in order to get insights from anyone in the organization. Another advantage of having these Fridays is that it allows for slow thinking, so that we are responding to our client's work instead of reacting to it.

We've been observing these mindful Fridays every week for the last seven years. For the last five years, our company has been named the best place to work. We've also done a well-above-average job of attracting and retaining top talent. We've been able to outthink the competition because we've prioritized the task of thinking.

STRATEGIES IN THE
COMMUNITY

Low voter turnout has been a problem in Baltimore City for years. Finally, local leaders decided to act. The first step was setting up a Facebook group for the city. Anyone from Baltimore City could join to talk about the city's most pressing problems. The group now has more than five thousand members. In the last mayoral election, voter turnout increased by close to forty percentage points.

Advances in social media alter the way we think about engaging communities and organizations. The old model had community leaders coming into churches and community centers and telling people what they should think, speaking to them as if they were children, much like the traditional atmosphere of the workplace. Here, technology enables citizens to sit with ideas and thoughts, so they can respond slowly, deliberately, and thoughtfully to potentially divisive issues. Instead of reacting like children, they are responding like adults. Everyone has a say, also like adults. When they feel as though they're connecting on a higher level, there is more of an incentive to stay engaged.

RHYTHM

Multitasking is not conducive to clear, innovative thinking. The head brain wants us to focus on one task at a time. Every time we switch to a minor task, like checking email, the head brain releases dopamine, a small reward for accomplishing an insignificant task. It's the equivalent of swallowing a teaspoon of sugar. Our brains like the dopamine and they want more, so the moment we return to our important work, the head brain, looking to score another hit, will encourage us to take on an additional minor task, like checking the basketball score. All this moving from one task to the other increases the stress hormone cortisol since the brain finds the constant starting and stopping exhausting. Studies show that multitaskers lose IQ points and do less thorough work.[CH4:N5]

We can all agree that the smartphone era has made the above phenomenon much worse. Getting over the addiction of multitasking—yes, it's an addiction—requires a change in behavior and routines. Fortunately, there are many tools available to help us maintain our focus.

The Productivity Planner is a daily planner. At the beginning of each day you write down your most important task for the coming day. Then, you list tasks of secondary importance. Finally, you note a third tier of tasks to be completed if you accomplish your first and second priorities. It also has a tool to track time so you can learn to work in shorter, more productive bursts. This tool is all about matching words with actions, so you spend time on activities that matter most.

The scheduling app Calendly helps you spend less time on things that don't matter. Instead of emails going up and back between two people as they try to schedule a meeting, one participant visits the Calendly page of the second participant and clicks on an available time. In less than thirty seconds, a meeting has been scheduled.

The average American spends 4.7 hours in front of the television every day. He or she also spends more than three hours a day on his or her smartphone. That's seven hours looking at a screen. The Moment app tracks how much you use your smartphone each day. One can set daily limits, or block out times for the entire family, so everyone around the dinner table will be engaged in the conversation, instead of sneaking peeks at their screens.

Trello, Evernote, and Slack all allow organizations to coordinate, collaborate, and communicate in an efficient and streamlined manner. In our own organization, Slack helped us reduce emails by 60 percent.

IS THIS
ENOUGH?

Engaging the brain in the head allows us to process information, notice patterns, and take away blind spots that prevent us from seeing a different future. Once engaged, we can begin innovating at work and in our own lives. Staying the course—especially when we face setbacks or when tediousness sets in—requires full engagement with our other two brains, in the heart and the gut.

FOR HELP ON HOW TO SET YOUR WEEKLY CADENCE AND RHYTHM TO SPEND THE MOST OF YOUR TIME WITH DELIBERATION AND ENGAGE YOUR FRONTAL CORTEX IN THE BEST POSSIBLE WAY, CHECK OUT *THE HOUR OF POWER* AT WWW.SHIFTTHEWORK.COM/TOOLS.

"Any fool can know. The point is to understand."
ALBERT EINSTEIN

DO THE WORK

EXPAND YOUR BRAIN

Visit TED Blog, and in the search bar type: 12 talks on understanding the brain. Or simply type this into your browser: https://blog.ted.com/12-talks-on-understanding-the-brain/

CHAPTER 5

DO YOU THUMP?

"I've learned that people will forget what you said, people will forget what you did, but people will never forget how you made them feel."

MAYA ANGELOU

We don't handle change well. We suck at making commitments, so we exchange our iPhones for new ones every year. We fail at the promises we make ourselves, which is why most people won't follow through on their New Year's resolutions. Instead of working on relationships, we choose to start over. We lack grit and determination. We aren't tenacious at those moments when we are struggling to reach the finish line.

THE BRAINS IN OUR HEADS CAN HELP PUT OUR THOUGHTS AND WORDS INTO ACTIONS.

Our head brain can notice patterns so we can design innovative game plans at work. The question is: What happens when we face adversity and the game plan looks like it's failing? How do we maintain a steady level of excitement, enthusiasm, and energy during these difficult periods? How do we have the courage to follow through when everyone is telling us it can't be done? How do we stand firm when people dismiss our dreams as childishly romantic?

Most of us begin new jobs with a sense of deep commitment. We want to make an impression on our bosses and new coworkers. We set out wanting to do an A+ job. If the boss asks for volunteers, our hands are the first to go up. Showing up early and staying a little late is fine because we're ALL IN. Somewhere along the way, this commitment and enthusiasm disappears for 70 percent of us. Soon, we are keeping our heads down at work and waiting for the clock to hit six. Returning home doesn't bring a sense of relief or elation. Instead, we feel as if we went through a battle. All we want to do is vegetate in front of the television. Maybe we'll decide to get a new job, but we won't have any sense of why the previous job failed, and the same pattern will follow us to our new position.

We now understand how to free ourselves from this mindset of rinse, replace, and repeat. Recent advances in neuroscience tell us that when the going gets tough, it's the brain in the heart that gets us going. It's why we pound our chests when making a guarantee to deliver on a promise. We never point to our heads to demonstrate the commitment is true.

The brain in the head has us checking off boxes, keeping us on a path, and sticking to a process. *How* you do it, however, is as important as *what* you do. The brain in the heart allows us to experience the moment and connect to our actions in a visceral manner. When we face adversity and aren't feeling great about our direction, it's the heart that steers us and decides whether the commitment is worth it.

We've been taught that the brain sends orders to the heart. This is true. What we haven't been taught is that the heart sends orders back to the brain. Research shows how signals sent from the heart impact our attention, perception, memory, and problem-solving. Think about moments when stress takes over your body. Your thinking is impaired. It's not the time we make our best decisions. It's why we tell people to take a deep breath before they decide on the next step to take.[CH5:N1]

It reminds me of a scene in the movie *Rudy*. The coach of the football team lays into one of his star players and says, "If you had a tenth of the heart of Ruettiger, you'd have made All-American by now!"

THE SEED SCHOOL AND THE
ONE THING I ALMOST FORGOT

When the SEED School of Maryland asked me to deliver a speech to 308 middle-school children, I felt flattered. Then, the nerves kicked in. In the past, I'd spoken to high-school students about my journey from Baltimore City kid to entrepreneur, but a gymnasium full of boisterous adolescents felt like a completely different challenge. Turns out, I was right.

Having always believed that giving back to the community is a privilege, I said yes. Plus, it's hard to think of a cause better than the SEED School of Maryland, a public, college-preparatory boarding school that provides tuition-free education to some of the state's most disadvantaged children. Admission is based on a lottery system, and for many of the students, it's a winning ticket that's the difference between life and death.

I pulled up to the fifty-two-acre campus on Baltimore City's west side. The grounds were as deserted as a dust-bowl town. A few minutes early, I went over, in my head, the themes my speech would cover: values, perseverance, leadership. Not exactly subjects that would strike a chord with a group of hyperactive, hormonal adolescents stuck in an assembly.

The anxiety intensified the moment I entered and saw bleachers on both sides of the enormous gym. It's hard enough to engage children when one is facing them. Now, I'd have to figure out how to position my body in a way that would allow me to make maximum eye contact with everyone in the audience. Also, the room was empty. I spent a couple of minutes wondering whether I messed up the time. I guess it's safe to say that I was looking for doubt anywhere I could find it.

Finally, the students started trickling into the room. Eventually, every seat was taken.

When the head of school quieted everyone down, he delivered a brief introduction and handed me the microphone. Any nerves I'd been experiencing immediately went away the second I started speaking. I felt relaxed, and the students were engaged as I opened with a humorous story from my youth.

Then, the microphone cut out. Suddenly, it was like I was whispering into a hurricane. People were yelling that they couldn't hear me. In a flash, I lost the audience. The children started talking and laughing with one another. Three students scrambled around me in an effort to solve the technical difficulty. It soon became clear that they didn't possess the expertise to deal with the issue.

For a second, I thought about bailing. Then, I considered the message it would send the children. I didn't want them thinking that quitting was the right response to dealing with setbacks. So, I proceeded to yell the speech at the top of my lungs. All eloquence and grace went right out the window, but the kids at both ends of the gym could hear me perfectly, or they had quieted down because I was yelling like a madman.

I proceeded to explain the importance of perseverance. I told them about growing up in Baltimore City, losing my mother at a young age, and going to the tough schools that they had narrowly escaped thanks to the SEED School. It's tough to sound inspiring when you're screaming at the top of your lungs, but I tried. I told them about attending Johns Hopkins, being an entrepreneur, and the constant effort to make my mark in my community and the world. Although relieved that I'd reached the end of my remarks, I felt discouraged by the thought that the students hadn't heard one word I said.

Ready to walk away, a teacher approached. She said some of the students had questions, a request that lifted my spirits. Perhaps they had heard my remarks.

The first child stepped to the microphone. He asked, "How much money do you make?"

After providing a diplomatic answer, I took a second question.

"What's it like being in business?"

In one minute—still yelling at the top of my lungs—I tried relaying to the group a sense of our company's day-to-day work.

Next, a young girl stepped up to the front and asked, "Do you think your mother is proud of you?"

I waited to answer, sensing she had something else to say.

She did.

"See, I lost my mother, too, at a young age, and I want to know if she's proud of me."

At that moment, I almost started bawling. Without a microphone, I'd managed to reach this girl. I'd shared something of value with her, and she responded with something of value. Mission accomplished, even if she were the only one in that entire gym who had heard my speech.

I LEARNED A VALUABLE LESSON THAT DAY: WHAT WE SAY MATTERS.

Sometimes we feel that nobody is listening. We haven't found the right words to express an idea, the microphone is broken, or the crowd is too busy on their phones to listen. If we're willing to put ourselves out there, we'll find people who're ready to find value in the message we want to share.

The head can devise all sorts of strategies for putting you in front of an audience, but it depends on the heart brain to make the connection.

"You never change things by fighting the existing reality. To change something, build a new model that makes the existing model obsolete."
R. BUCKMINSTER FULLER

WE'RE WAITING FOR YOU

If you courageously assumed the mantle of influencer and shared your most important message with us, what would it be?

WRITE IT DOWN BELOW.

....

....

....

....

....

Share your message above with at least one stranger a day for the next seven days. Be brave; be bold. Make a shift!

TONY ROBBINS'S

SIX NEEDS

SIX NEEDS DRIVE HUMAN BEHAVIOR, ACCORDING TO TONY ROBBINS:

1. CERTAINTY/COMFORT

2. UNCERTAINTY/VARIETY

3. SIGNIFICANCE

4. LOVE AND CONNECTION

5. GROWTH

6. CONTRIBUTION[CH5:N2]

What do the first four have in common? They're what we'd call matters of the heart. Number 4, "Love and Connection," is particularly crucial. It brings to mind the Harvard happiness study discussed in chapter 3, which identified connection as the number-one factor in determining a person's overall satisfaction with life.

Our thirst for connectivity is undeniable. It's why we check our phones every five minutes and feel a need to share every like and interest on Facebook. It's the reason we prefer to stay in a job that is familiar even if it brings us no satisfaction. An insufficient connection, we mistakenly believe, is better than no connection at all. Some people turn to drugs and alcohol in trying to fill this need. Someone taking drugs doesn't know if he will live or die. What he does know is that he will feel something.

It will make him feel important, and it's a lifestyle that connects the user to a group of people with the same limited interests.

The challenge for all of us is to fill the need of connection in a constructive, rather than destructive, manner. Unfortunately, the destructive path, as one might imagine, is far easier to pursue. Fixing your marriage, pursuing an education, and finding engaging work is much more difficult than doing drugs, joining a gang, or even staying in the same dead-end job.

WHAT THE NEUROSCIENCE OF THE

HEART TELLS US ABOUT ENGAGEMENT

THE HEART HAS ENERGY. We all know someone with infectious energy. What we don't all know is the biological explanation for how people can change the mood in a room.

The heart acts like a magnetic field, radiating its moods to anyone in proximity. When we engage someone, whether through touch or proximity, our hearts transfer an energy forward. While our bodies absorb the mood, our heart brain— working with our head brain—releases an energy back at the person.

Think of the brain in the heart as the body's power plant. Power plants, as we know, don't just use energy; they generate it. The heart brain manufactures energy through its connection to the head brain. Consider that the brain in the heart has forty thousand neurons, a fraction of the 86 billion neurons located in the brain in the head.[CH5:N3] Still, the small cluster of neurons acts as a key portal of the neural network, pulling thoughts and ideas from the brain in the head, and releasing the subsequent

energy into the world. The head brain will notice if the person standing across from you is happy, but it's only once the heart brain receives the message that it releases the oxytocin and pumps the blood at a faster rate. It will change the way the muscles in our face move. Our voices will sound different. The person standing across from you will observe and absorb these changes.

THE HEART COMMUNICATES. We've all seen our share of Oscar-worthy performances. The actor can use the head brain to make the choice to use a certain accent or move his or her body in a way that properly reflects the character. Intellectualizing the role will only carry the actor so far. The magic happens when the actor can put him or herself in the emotional space of the character he or she is playing. The heart has to be pumping at the right rate that will allow the voice to crack just perfectly right before breaking down in tears, or it won't be believable.

THE HEART REMEMBERS. The heart sends as many, if not more, messages to the head brain than it receives. Researchers in the field of energy cardiology have found that nearly all heart-transplant recipients report experiencing memories and emotional responses that appear to have come from the donor's personality. What this means is that the heart creates thinking hormones similar to the type created in the head brain.

In her book, *A Change of Heart: A Memoir*, Claire Sylvia shares her experience after receiving the heart and lungs of her donor—a teenaged boy who died in a motorcycle accident— describing a change in her cravings, behaviors, and emotions. She reportedly acquired her donor's love for beer and chicken nuggets. Sylvia also found she became more aggressive and impulsive. After she sought out the family of her donor, these physical and psychological changes were confirmed. Sylvia's story is only one of a number of documented case studies that supports that the heart maintains our memories.

THE HEART TASTES. The human genome has twenty-five bitter taste receptors, twelve of which, according to The School of Biomedical Sciences team at the University of Queensland, are located in the heart. As part of this team's ongoing research into the growth of human hearts during disease, they found that when the taste receptors are activated with a chemical—that we taste as bitter—the contractile function of the heart was almost completely inhibited.

THE HEART BREAKS. Even the healthiest of individuals can experience a broken heart. Broken heart syndrome, also called stress-induced cardiomyopathy, is the body's reaction to stressful moments most often linked to our relationships. CH5:N4 The death of a loved one, a breakup, physical separation, betrayal—many of us have, or will, experience some kind of heartbreak in our lifetime. The brain in the heart detects the surge of stress hormones and responds to these heart-wrenching moments with ache—physical and emotional—but in most cases promises to heal with time. The heart craves connection and engagement, and when that connection is diminished or broken, our body responds.

THE HEART HEARS. A great song plays on the radio, and we feel it in our hearts. A musical theorist can explain why certain notes played together can sound particularly pleasing, but we feel the greatness in our hearts, not in our heads. It's our hearts that beat faster when the song is recognized by our ears.

IT'S NOT WHAT YOU SAY,
BUT HOW YOU SAY IT

In the 1950s, Dr. Albert Mehrabian studied the elements that form the basis of communication. He determined that only 7 percent of communication is about the actual words spoken. In fact, 38 percent is about voice quality—pitch, tone, volume, rhythm, and frequency. The most significant element of communication is body language. It accounts for 55 percent of communication. Overall, 93 percent of communication isn't about what you say, but how you say it.

Try this exercise. Repeat the following sentences, while emphasizing the italicized words. You will notice that although the words are the same, each sentence has a completely different meaning.

I didn't steal her money. (It wasn't me.)

I didn't *steal* her money.
(I took money from her, but I don't consider it stolen.)

I didn't steal *her* money.
(I stole money, but not from her.)

I didn't steal her *money*.
(I took something from her, but it wasn't money.)

In the workplace, we believe email is the fastest, most efficient system to communicate. The reality is that it's often difficult to interpret the sender's intention. Was the person genuinely angry, or speaking sarcastically? It's symbolic of the workplace environments we've created. We leave no room for feelings and connecting with our coworkers and the work at hand. It's difficult to work with purpose when feelings are suppressed throughout the day.

DO THE WORK

TO EMAIL OR NOT TO EMAIL?

Is an emotionally charged topic on your mind, with a high probability of being taken the wrong way? Send the person a text message, asking when they have a few minutes to talk on the phone. Better yet, make arrangements to have the conversation in person.

After you have the opportunity to give this a try, revisit this page and record the outcome.

SHIFT THE WORK

*"The single biggest problem in communication is
the illusion that it has taken place."*
GEORGE BERNARD SHAW

The number-two TED Talk of all time is not about curing cancer,
or how 3-D printing will change the world. It's "Your Body
Language May Shape Who You Are," by Amy Cuddy. We all
crave powerful connection.

GENERATION XERS ARE

ALL ABOUT HEART

The baby boomer generation was all about compliance, control,
and compensation. People were content with being part of a
hierarchical system. They devoted themselves to the company
and the job. They were a generation in tune with the brain in the
head, which is about planning, innovation, and following a path.

The members of Generation X took a different path. They
believed in following their passion. It's the generation that grew
up on *Rocky* and *Rambo*. Generation Xers believe that where
there is a will there's a way, and nobody should ever accept
that no is the end of the road—a sense of determination that
perfectly aligns with the brain in the heart.

OUR HEARTS IN THE WORKPLACE

Level of passion speaks to the question of whether you own what you do in your heart. Does your belief in the company make you want to do a great job at work, or are you completely indifferent to the company's success? We surveyed a group of employees to learn whether, if knowing what they now know, they would accept the job again if offered it today—72 percent answered in the affirmative. This sounds high, but is it really? Over a quarter of the company would choose not to take the job again. Would you go to a restaurant if a quarter of the diners said they'd never come back? It's no wonder we have such tremendous divisions in this country. People aren't finding an outlet for their passion at work, so at home they turn to politics to fill the need. It's negative passion that results in blaming others and pointing fingers for the great frustrations in our lives. When a kid isn't being paid attention to, he or she acts out, essentially what the adult workforce is doing right now.

Willingness to suffer is the question of whether you're prepared to push through when confronted with the harder, less appealing aspects of the job. Not every task we are asked to do at work is glamorous or fulfilling, but if we believe in the overall mission, we are willing to tolerate these elements as part of the job. It's when we don't agree with the mission that they become onerous.

Level of resiliency is about whether we'll break before we see something through to the end. We all hold up commitment as a virtue, and none of us wants to live with regret, but it's easier said than done, especially when we are trapped in a negative situation. Say you decide to do a triathlon but only spend one month training. Midway through the race, you're forced to drop out. You won't walk away feeling satisfied that you gave it your all. Further, the fact that you wasted little time training will not provide comfort. All you'll feel is regret over not having made a greater commitment to succeeding and having wasted time on a half-assed effort. Failing should only feel okay when we know we've exhausted all options. When we're indifferent to it, it means it's time for serious reflection.

STRATEGIES FOR YOU:

MOTION CREATES EMOTION

Annie O'Dell (nicknamed, AOD) exuded passion. A gentle soul with an animated spirit, Annie finished college in her home state of Minnesota and drove to Washington, DC, to take a job at a company she knew nothing about. Three months later, she left that job to work with our company. We gave her the position of project specialist, meaning she did a little of everything. In twenty-five years of working in various businesses, I've yet to encounter anyone who's matched the level of enthusiasm Annie brought to her job at SHIFT.

Right away, she stood out for having a unique point of view, one that was shaped by her connection with the brain in her heart. We asked her to put together a thank-you gift for our clients. In years past, we took a pretty standard approach to this gesture—fruit baskets, gift cards to restaurants, holiday cards. That year we settled on the idea of bringing a catered lunch into the offices of our clients. Lunch in the middle of the week, Annie thought, was a nice gesture, but it too was ordinary. Ice cream sundaes for dessert on a Wednesday afternoon, on the other hand, would prove that we put serious thought into the gift. Suffice it to say, the clients were blown away. They commented on the utter joy as, like children, they topped their sundaes with fudge and sprinkles in the middle of a workday.

At one point, Annie noticed that in company meetings, everyone was dividing their attention between the proceedings and the phones in their hands. She took a trash can, covered it with green felt, decorated the lid like a frog face, and called it Kevin. As a new rule moving forward, everyone would deposit his or her phone into Kevin at the start of meetings. She could have suggested we all put our phones into a nondescript, cardboard box, or leave them in our office, but like with the ice cream sundaes, she had a way of electrifying people's hearts and getting them excited about ideas.

Soon after convincing several members of the team to jump out of an airplane for the first time, she felt a pain in her stomach. Doctors discovered that she had a rare form of sarcoma. For two years, she met the cancer with the grace of an angel.

Her sickness came right in the middle of 2009, as our company struggled during the recession. Annie wasn't physically present at work during this time, but I'd be lying if I said I wasn't channeling her spirit and style as I tried to dig our company out of the ditch.

She came out from Minnesota to visit us one last time. To this day, I'm haunted by the memory of her telling us that she could feel the cancer inside her body. Three weeks later, she asked us to visit her in order to say goodbye. There she was in her hometown, surrounded by loved ones. Meeting her family and the people from her hometown, it was clear that this was someone who touched everyone deeply, as we suspected. Even though it was tragically early, she said she had no regrets.

My wife, Erica, was pregnant at the time, and we were having a debate over what to name the child. For a reason that wouldn't become clear to me until much later, I decided to ask this woman who was close to death what we should name this new life. Annie picked Eliana (and we would call her Ellie for short); the name sounded right coming out of her mouth. Then, she told me to remember that love always prevails and goodness never dies.

Several years ago, I had the privilege of interviewing Jack Welch about his new book The Real-Life MBA. I was overcome as he began to tell me about a career assessment process mentioned in the book that he called Area of Destiny. AOD. On one axis is a person's passion. The other axis is the person's skills. The idea is to identify where passions and skills meet. Annie, our AOD, was definitely in the top right quadrant. Her passion perfectly aligned with her skills because her skill was her passion. She had this natural ability to light people up, whether it was her colleagues, clients, friends, or complete strangers. A more appropriate job title for her would have been "director of inspiration."

"You can't read the label on the jar you're in."
HUGH MACLEOD OF GAPINGVOID ART

BOOSTING YOUR B'S

Garrett White, author of Wake Up Warrior, believes that breaking down our lives into two dimensions—personal and business—is simplistic and outdated. Instead, he identifies four elements of our lives that must be addressed in order to get the most out of our lives:

☐ **THE BODY (HEALTH)**

☐ **THE BEING (SPIRITUALITY)**

☐ **THE BALANCE (RELATIONSHIPS)**

☐ **THE BUSINESS (DOUBLING PROFITS WHILE WORKING 40 PERCENT LESS)**

Rate your performance on a scale of 1 to 10 (1 being *at risk*, 5-6 being *moderate*, 10 being *high*) in all four of these elements. If you are ranking yourself below a 5 in any of these categories, ask yourself what you can start doing today to shift those numbers.

CONNECT WITH YOUR HEART

We've discussed the importance of journaling in the morning. The company that created *The Productivity Planner* also has a product called *The Five Minute Journal: A Happier You in Five Minutes a Day*. Again, it's the idea of shift your language, shift your mindset. Through using specific language to write down

SHIFT THE WORK

our goals, we can better connect to the heart brain. Approach the coming day with a mindset of gratitude. Instead of what we "have to" do, we write down what we "get to" do.

Normally, commitment to exercise is a product of the head brain. We know it's healthy, and we devise a plan to achieve certain goals. The popular exercise program CrossFit, on the other hand, is about connecting people to fitness and health through the brain in the heart. Participants see themselves as members of a community. The commitment one makes to oneself comes through the bonds formed with the other members. People visit one of the "boxes," or affiliated gyms, unsure of what exercises they'll be doing that day. The hour starts with an instructional period and is followed by group stretching. Finally, everyone goes through the WOD (workout of the day) together, and each individual's results are put up on a board. It's no coincidence that if you look at the people in your life who do CrossFit, they're probably the people you find most passionate about living.

STRATEGIES:

WORKPLACE

What if your organization established an environment where workers felt connected to each other and the work? It starts with putting the values and vision of the company into words, so there is a clear picture of how a worker will experience those values on a daily basis. This manifesto should be thoughtful about the full life cycle of an employee, from the time the worker comes in for the interview through the person's departure. Think of how nice it would be to work for a company that acknowledges that there will be an end of the road. After all, not everyone will become an executive. People will top out at a certain point and will want to pursue a different challenge. Most companies act angry when a worker decides to depart, even though they've been doing the same job for ten years without any promotion. A company that acknowledges the end at the beginning is more likely not only to help workers achieve their goals but also to have workers who are more engaged during their time with the organization.

Take the example of the consulting firm McKinsey. Upward of 75 percent of senior partners leave to become corporate executives. McKinsey accepts these departures as a badge of honor, a sign that the company's culture produces some of the best in the field. McKinsey accepts that employees view the company as a stepping stone, and because of this reputation they're able to recruit top-level talent who will work their tails off to get that next great job.CH5:N6

Too often, companies see falling profits and blame it on product or distribution issues. They never think to look at the issue of engagement, even though we know a disengaged worker is a less productive worker. Glassdoor has emerged as the Yelp for businesses. It's a site workers can visit to see whether potential employers understand what employees of the company are thinking and feeling at work. Read through enough reviews of a company, and you'll be able to quickly tell whether the workers are engaged or not. It's unfortunate that a majority of company leaders aren't constantly assessing whether workers feel connected to the energy of the company. Want to see how we dealt with a tough Glassdoor review? Watch the video at www. shiftthework.com/tools.

At SHIFT, we care deeply about maintaining the passion of our workers, which is why we grant employees month-long, sponsored sabbaticals after ten years at the company.

COMMUNITY MISSION

When you can see it, you can feel it. That's empathy.

Growing up in Baltimore City, "on the wrong side of the tracks," it was easy for me to be anything but a fan of the police. That limited point of view changed when I arrived at Hopkins and a criminal-justice professor required everyone in the class to participate in a police ride-along.

They paired me up with officers attached to the northern district of Baltimore City, an area that included the dangerous North Avenue. This was in the late nineties, a particularly violent period in our city. It was a scary adventure, even for someone like me who'd grown up in a similarly rough neighborhood.

We spent the first half of the night tracking down carjackers, sending loitering children home, and arresting vagrants for disorderly conduct. It brought back a lot of memories and feelings, although I'd have to keep reminding myself that I was now playing for the other side. Slowly, as the night wore on, I began to appreciate the difficulty of the officers' jobs. They were forced to make split-second decisions and defuse potentially violent situations. Present in all their interactions was the constant struggle to maintain the difficult balance between firmness and civility.

At around midnight, as we took a short coffee break in a convenience store lot, the dispatcher interrupted with a call that there'd been a shooting on the Hopkins campus. I'd always thought of my campus as a bubble, impervious to the violence that surrounded it. It even had its own security force, but there we were flipping on the siren and racing to the scene. When we arrived at the campus entrance, the officers grew agitated, as they struggled to find a clear route to the library, where the shooting had taken place. This part of the city wasn't a normal route on their beat. I leaned forward from the back seat and, through the glass divider, began directing them to the scene.

Moments later, we arrived, driving onto the grass. Blood was everywhere. The ambulance hadn't even arrived yet. The victim was on the ground, a bullet wound to the head. Bystanders cared for him, but he was dead. I'd seen violence up close before, but I thought I'd left it all behind when I left my old neighborhood for life at Hopkins.

The officers asked if anyone saw the shooter. A witness pointed in the direction of a dormitory. The officer took off, and I followed close behind. By the time we arrived, Hopkins had already detained the man. They handed him to the officer, and the three of us began walking back to the car, shoulder to shoulder.

"Hey," the shooter said to me, looking me dead in the eye. "Do you have the time?"

My officers were tasked with accompanying the corpse to the emergency room. As the doctors worked on the man, the officers went through his wallet, trying to look for identification. I'm the one who had to tell them that all of the IDs were fake. Meanwhile, the whole time I'm thinking that once they figured out the name, they'd be the ones who'd have to drive to the person's house and inform the next of kin. Wow.

WHEN'S THE LAST TIME YOU WALKED IN SOMEONE ELSE'S SHOES?

It's not enough to read about experiences in a book or listen to a description of what a person goes through. Classrooms only take you so far. You won't know it, until you live it. It's like when my wife goes away for the weekend and leaves me in charge of the kids. Only then do I truly appreciate the difficulty of the work she does every single day.

"You may say I'm a dreamer, but I'm not the only one. I hope someday you'll join us. And the world will live as one."
JOHN LENNON

DO THE WORK

YOUR NEXT RIDE ALONG.

Who in your life do you have a hard time relating to right now because you believe they're slacking or showing a lack of ingenuity?

WRITE THEIR NAME HERE.

Find a way to live in this person's shoes for a day to experience the world through their eyes. Then, try to help them make a shift.

The heart brain is literally an energy field. If we want to connect with people, we need to stand where they are standing, to see the world through their hearts. Engagement doesn't mean dropping a check to a food bank into a mailbox. It requires taking the time to volunteer at the food bank, seeing the look on the recipients' faces as they pick up the food. It means taking the opportunity to sleep at a homeless shelter. When was the last time you spoke to the person who cleans your office at night? This doesn't mean you need to scrub toilets alongside the person, but try to get a sense of what their life is like. If people had a sense of what it's like to work at a DMV, they'd be less quick to harshly judge its workers. Maybe you'd discover that they think we are lazy, impatient, and ignorant. Use your lunch break to spend time with workers from a different department. If you're an executive, then spend time with a lower-level employee. Lower-level employees should try to get a sense of what life is like for the people in charge. If you drive to work every day, try taking the bus or subway. What's important is that you embrace these opportunities without an agenda besides deepening the connection with the other person and the world around you.

Several years ago, I had the privilege of visiting Detroit's Superhero Training Academy, a not-for-profit that empowers children to tap into their inner superheroes and unleash their potential. As part of the curriculum, the children dress up as superheroes, wearing masks and capes, and go through the neighborhood executing various challenges. The students even choose superhero names. The purpose of this dress-up is to place them in the emotional space of the person, or superhero, that they want to become.

WORLDS OF OTHER PEOPLE

It's easy to criticize other people's words and actions. When we take the time to stand in their world, we can appreciate that they face their own struggles and challenges. Everything becomes relative. Suddenly, our empathy, patience, and tolerance are increased.

In college, I was playing football and struggling through my classes, all while taking care of my mother. I was also running several businesses, one of them a house-painting company. One day, I had to be at two places at one time. A lot of money was on the line, and I needed someone to run an errand for me, so I could meet a potential client. I made some calls to friends and partners, but nobody was available. Finally, I turned to my roommate and football teammate, a kid who was busy smoking and drinking his way through school.

I explained to him that I was in a bind, and I'd truly appreciate it if he could pick up an order from the paint store for a project that was scheduled to start later that day. I told him I'd pay him for his time and would even throw in dinner and a six-pack.

This roommate, who woke up at two in the afternoon every day, looked up at me and said, "No, I'm really busy today."

At the time, I was quite pissed at him and ended up not talking to him for weeks. Years later, a different perspective leads me to a different, more charitable response. The stress my roommate felt was real to him. There must've been something going on in his head and life, some type of pressure or anxiety, that made him want to drink and smoke all day. He was genuinely stressed out, and the task I was throwing at him probably did feel like more than he could handle.

Engagement is about connecting through the heart brain and absorbing the energy that is emitted from the people around us. We live in a society that tells people it's okay to be self-centered and constantly worried about our own feelings. This doesn't make people happier. If anything, it closes us off and prevents us from trying to engage other people. Nobody stops to consider that we'd all be happier if we took the initiative to rise above our current situation and tap into the worlds of other people. If we can become the coach, the person who manages the flow of the surrounding energy, we can learn things about ourselves from others. These are the connections that would truly make a difference in making us more engaged with family, our community, and ourselves.

BEING BLIND

FOR AN AFTERNOON

With our corporate clients, experience has taught us that a change of perspective can lead to a change in perception. The National Federation of the Blind, which is located in Baltimore, has an exercise where they blindfold visitors for three hours and challenge them to handle everyday tasks. After about thirty minutes of suffering through this sensory deprivation, the other senses kick in and become amplified. Suddenly, you can hear everything. Objects feel differently when you touch them. You can sense when people pass in front of you or are approaching from behind.

The experience allows one to appreciate not only how blind people manage to get through life, but also how a musician like Beethoven used his handicap to become one of the greatest composers of all time. It's no coincidence that some of the great entrepreneurs are learning disabled. The process of overcoming the obstacle and finding a workaround has allowed them to discover fresh ways of looking at how we live.

BEING BLIND

FOR AN EVENING

In 2016, for our company holiday party, we put on an evening called "Dining in the Dark." Executives and entrepreneurs from Baltimore and Boston were invited to an elegant dinner. They were broken into groups of six, then blindfolded before taking their seats. Unbeknownst to them, at select tables sat people who recently reentered society after serving time in prison. One was a woman who had her first child at thirteen years of age. Another was a man who had spent twenty years in prison on a wrongful conviction. A second woman was trying to get her GED at the age of thirty-two. As dinner was served, the participants went around the table answering three questions we had prepared in advance. What has been their greatest life struggle? What are the greatest issues facing Baltimore? What are their hopes for Baltimore?

The answers provided by the recently freed guests didn't surprise me. These were my neighbors growing up. (The woman who had the baby at thirteen attended the same high school as me.) Their stories of finding themselves stuck on a destructive, desolate path were standard for that community, which I knew well.

The entrepreneurs and executives were part of the world I now inhabited.

Hearing these stories could inspire empathy, but would our business leaders be called to do something about it?

We performed a mini social experiment. We asked our guests to raise a hand if they met someone whose story amazed them. Hands went up in the air. We then shared with them the struggle these people face when looking for work. Would the people in the room, we asked, hire this amazing person they met? If so, raise a hand. Every single person raised a hand, and we instructed them to remove their blindfolds. Watch the video at www.shiftthework.com/tools.

The executives and entrepreneurs in that room connected to these less fortunate people through a shared humanity. Empathy is a drug. It releases dopamine in the brain, oxytocin in the heart, and serotonin in the gut. This biochemical reaction is the body's way of connecting us to the greater good and prompting us to act on that feeling. It's the body's way of asking, "What are you going to do about this situation?" Letting the opportunity slip away creates a void. Don't think your body won't take note if you ignore the call.

After 9/11, New York City came up with the mantra, "If you see something, say something." If your body is telling you something is off, the proper response is not to sit back and stay quiet. The obligation is to do everything you can to improve the situation. Likewise, if a coworker is taking an action that is inconsistent with the company's stated values, even if the person is senior and you are low man on the totem pole, it is incumbent on you to take a stand.

TOOLS, TECHNOLOGY, AND TRAINING

In the backs of our minds, we know we get only one go-around, and we don't know when it will end. What if we could more often operate based on this reality? If we were continually aware of our fragile state, we'd be our best selves every day, our hearts to lead us. Life begins with our relationships. Life isn't celebrated by consumption, but through connections. It's about the energy we give and take.

THE WINS

One of the major reasons people leave their jobs is because they don't feel as if they're making progress and growing. What if your organization celebrated even the small victories of its workers? This could mean ringing a bell when a deal is closed, or banging a gong when a major project is completed. Some companies will use a weekly newsletter to highlight the achievements of employees.

RIFF

In business, people are told not to discuss their deepest feelings and insecurities. What ends up happening is that workers carry around their resentments, disappointments, and frustrations. With no outlet for them, the feelings build up over time, and the workers grow more and more bitter by the day.

SHIFT THE WORK

THE NAME-TAG EXERCISE.

At your next company meeting, have everyone wear a name tag. When the meeting is over, collect the name tags in a pile. Have each participant randomly select a name tag. On the back of the name tag, each person should write down what they appreciate about the person whose name they've drawn. What gifts does the person possess? What makes them an asset to the organization?

Recently, I decided I wanted to address this problem by connecting with our team in a uniquely different way. Also, a majority of my work is done outside the office, and I feared that this contributed to the team's feeling disconnected from me emotionally. To deepen the level of engagement, I set up a channel on Slack called #CEOriff. Every day, I send the team a message. Sometimes I riff about what is happening in the company. I include both good and bad news. What are the patterns we want to continue and reinforce? What are the patterns we want to shift? It could be me singing the praises of a team member who responded to a client problem in a particularly clever manner. I'm particularly interested in highlighting models of behavior. Other times, I'll introduce a general business lesson or thoughts on an interesting article I came across. I've even riffed about my own life. I'll tell them about how my family spent the weekend, or a time I disappointed a loved one. I once riffed on my daughter calling me out for not being present and spending too much time on my phone.

Sometimes the riff will take one minute to watch. Other times, it will take ten minutes. In the nine months since launching this initiative, I've missed only one day.

Most importantly, the riff gives everyone at the company an indication of how I experience certain events. It proves to them that the big boss can respond emotionally, and not just intellectually. It's a glimpse into my heart brain. If I can get my team to know me on a level that is separate from the head brain, then it frees them up to share their authentic selves, too. Watch a sample CEO riff at www.shiftthework.com/tools.

15FIVE

15Five is software designed to facilitate constructive conversations between employees and management. Employees take fifteen minutes at the beginning of the day to draft their thoughts on everything from the status of current projects and feelings about the work culture to priorities and challenges moving forward. Companies can customize the fields and questions, so the questionnaire is relevant to specific departments and workers. It should take a manager no longer than five minutes to review the feedback.

The point of the software is to create a culture of feedback. It creates set points, so managers and employees can see if progress has been made on various fronts. In any organization, workers need to feel connected to the organization's broader mission, and this is partially accomplished when they feel as if their concerns and feelings are being heard and addressed.

INSIGHTS

The TTI Success Insights assessment measures a worker's behavior, motivations, emotional intelligence, acumen, and skills, so he or she can exploit those talents. Objectively measuring our personalities and skills allows us to determine what kind of work makes our hearts beat faster.

GRATITUDE

LETTERS

At SHIFT we enacted an exercise called "The Gratitude Box of Letters." Employees are challenged to send out a letter of gratitude, every week, for an entire year. The recipient can be either a professional or personal contact. A letter may thank a business client for opening the letter writer's eyes to something amazing in the world, or it may thank a former teacher who taught the person a valuable lesson many years ago.

CAN YOU GET TO TWENTY-FIVE? Write down twenty-five reasons why you go to work every morning. It could be that you want to pay off your mortgage, you want a job that makes even a small, positive difference in the world, or you're seeking a professional challenge. Go through the list and ask yourself whether your current job satisfies these reasons.

THIRD BRAIN

Using our head brains to innovate and our heart brains to inspire isn't good enough to achieve different results if we aren't also connected, behaving like a hive, with a collective consciousness. This can happen only when we grasp the importance of the third piece of the puzzle, the brain in the gut: the intelligence center that has us thinking not for ourselves, but for the rest of the world.

CHAPTER 6

NOT GUTS, NO GLORY!

"Be the change you want to see in others."
MAHATMA GANDHI

HAVING THE GUTS TO LEAN IN

Readers of this book might assume that every action in my life is directed by a sense of mission, that every time I open my mouth, put my fingers to the keyboard, or speak to a client, I'm taking into consideration the impact it will have on the community and the world. In reality, I struggle with living up to my ideals every single day. Ideally, I'm always looking for ways to integrate my values into my life, but I, like everyone else, have problems with managing my priorities and remembering what's important in life.

Five years ago, I thought long and hard about my obligations toward the city I love. It was time to act on my principles. Politics seemed like a good entry point, so I applied to be on the Baltimore City school board. The universe didn't agree with my plans, and my application was swiftly denied. With that setback, I gave up any desire to get involved in local politics on a formal level. In the coming years, I'd channel my desire to make an impact by helping companies, organizations, and entrepreneurs integrate broader missions into their work culture.

In the spring of 2016, twenty-five-year-old Freddie Gray was arrested by Baltimore police, and later died in the hospital. The public reacted with outrage, and a lawyer hired by the family agrees that the arrest was unjust, saying, "Running while black is not probable cause." After the Freddie Gray episode, the Baltimore City school board chairman approached SHIFT asking for help. Coincidentally, this is the man who'd been selected over me for the position I coveted years before (they made the right call!). He wanted SHIFT's support with interviewing and assessing candidates for the position of Baltimore City school board CEO.

My wife, who works in PR, sized up the opportunity as a no-win situation. If the eventual CEO worked out, then nobody would care that we played a role in the selection process. On the other hand, if the recommended candidate ended up being a disaster, the press would surely come knocking when looking for someone to blame. They'd criticize placing the process in the hands of a private corporation, and we'd be held up as an example of greed ruining the public school system. This seemed like the more probable outcome given that the average tenure of a city superintendent in an urban environment is less than two years. Being publicly blamed for a major failure of the public school system wouldn't be good for business. Logic was suggesting we lean out.

But civic duty called, and we decided to lean in. We made the vetting process as transparent as humanly possible. Ultimately, the board independently selected the candidate we recommended. Still, it didn't take long to catch flack from the press. Immediately, there were articles about the "secretive" and "underhanded" process to select the new board CEO. To the critics, it didn't matter that we were putting our best foot forward and acting with the city's interest in mind. We'd worked long and hard to build a solid reputation, and it sucked that this one episode threatened to damage our integrity.

The episode shifted my perspective on politicians. Sure, there are plenty of politicians who act out of corporate or self-interests. Yet, there is still a considerable group of other politicians who want to make a difference and help their communities, and they put themselves in the public eye knowing they, their families, their good work, and their reputations will take a good amount of abuse.

In the end, we decided against issuing a comment. We weren't going to acknowledge that there was even a question of impropriety. We knew our actions were pure and just.

If presented with the same opportunity, our team would make the same choice again without hesitation. In life, there's a constant battle between what you *should* do and what you *need* to do. We *shouldn't* have taken on this job. Something inside us told us we *needed* to lean in. It was a biological need that we couldn't ignore, with the brains in our guts calling us to a higher purpose. That's why we didn't feel regret.

REGRET ONLY SURFACES IN THOSE MOMENTS WHEN YOU IGNORE THE *NEED* AND INSTEAD FOLLOW THE *SHOULD*.

"If something is important enough, even if the odds are against you, you should still do it."

ELON MUSK

DO THE WORK

SHOULDA, COULDA

WHAT ACTION DO YOU KNOW YOU NEED TO TAKE, BUT HAVE JUSTIFIED YOU SHOULDN'T?

Write it below, and then write the three biggest reasons why you NEED to make it happen

The role of the gut brain in calling us to a higher purpose is best summed up by the famous quote from *The Three Musketeers,* "All for one and one for all, united we stand divided we fall."

We know what it looks like to lead a moral and ethical life, but leading our lives according to those ideals is difficult. How many times have we seen another country treat its homeless and disabled in a way we find appalling, yet, on a daily basis, we find ourselves driving past these same people without lifting a finger? As a nation, we are charitable, but we're also regularly missing opportunities to contribute more or make charity a routine part of our lives.

Is closing our eyes to the half of the country whose lives are heading in the wrong direction morally acceptable simply because our lives may be headed in the right direction? We should be thankful if our lives are going well, but it doesn't give us a free pass to ignore the misery that others experience.

The movies we love are the ones where the hero puts his own life at risk to save many. The selflessness of soldiers, firefighters, and police is the reason we celebrate them in our society. In the book *Deep Survival: Who Lives, Who Dies, and Why*, Laurence Gonzales uses science and anecdotes to show that the people who survive life's great challenges are not the ones who act out of self-interest. Rather, it's the people who are focused on achieving the best result for the entire group.

THE NEUROSCIENCE OF
THE GUT

The bears have stopped chasing us through the wilderness, yet we live in an environment that requires an ability to cope with a much higher level of stress. We're taxing not only our physical muscles but also our emotional, mental, and spiritual ones. We rely on the gut to manage these greater levels of anxiety. Ninety-five percent of serotonin, the neurotransmitter that keeps our mood balanced and our thinking clear, is produced in the gut. In addition, 100 million neurons sit in the walls of our gut, which is more than in the spinal cord or the peripheral nervous system. CH6:N1

Nestled in the center of the gut is 90 percent of our immune system, the part of our body that keeps us healthy.CH6:N2 The American Gut Project focuses on researching fecal matter to gain a greater understanding of the connection between the microbiome and health. Anyone can send them a sample of their feces, and they will test it for ailments and diseases. The goal is that in the future doctors will be able to look at someone's fecal matter and tell whether the person has cancer, is in need of a certain vitamin, or is dehydrated.

The gut is connected to the other intelligence centers in the body. Information is relayed from the gut brain to the head brain via the vagus nerve, the longest nerve in the body. For the most part, this connection is one-sided, with almost all of the information going from the gut to the head. During sleep, it's the gut brain that processes information. When people talk about getting butterflies before a speech, they are talking about cortisol secreting from the adrenal glands. This is the stress hormone, and 70 percent of our body's cortisol is secreted from this location. The head brain is processing the situation, but it's the gut brain that's taking on the primary role of responding.

"YOU ARE WHAT YOU EAT." The gut is an ecosystem. Dr. Mark Gordon, a renowned researcher, directs the Washington University School of Medicine's Center for Genome Sciences and Systems Biology. He has been studying the microbiome for decades. Much of this research is just starting

IS YOUR GUT FIT? At a recent conference, I saw Naveen Jain speak about his new company, Viome— founded on the premise of creating a world where illness is elective. He talked about the simple, yet profound idea of taking control of your health, and becoming the CEO of your wellness.

Your gut has a unique army of approximately 40 trillion microorganisms that, when functioning in tune with your body's ecosystem, can maximize your health and potentially prevent disease. Learn more about their science and how it can help you with your gut health at www.viome.com.

to hit mainstream behavior psychologists to determine how gut health—specifically the bacteria that take up residence in the gut—defines who we become.

ALL ABOUT THE GUT

Millennials, with their need to impact the collective, are more connected to the brain in the gut than any generation before them. This sensitivity to what's happening in the gut brain and being in tune with all three brains makes them ideal candidates for tackling the world's greatest problems and becoming the greatest generation. People of older generations like to put down millennials as soft, claiming they know little about life because they've never had to endure the hardships of earlier generations. There's no question that adversity is instructive, but does this mean we should make the lives of millennials harder just to teach them a lesson?

We should be thankful that the people of this generation are faced with fewer adversities. This freedom allows them to engage the workplace in a unique manner. They don't have to worry about their paycheck, meaning they are free to focus on solving world problems like climate change, disease, hunger, racism, and inequality. It's our good fortune that they seem up for the challenge. What some people see as laziness is really a refusal to settle for any work that isn't meaningful. Millennials would rather work as baristas than take a job that isn't socially conscious. They're willing to live in their childhood bedrooms until they find the right job, by which they mean one that has a mission they can get behind. As consumers with disposable incomes, they're speaking loud and clear. They'll buy TOMS shoes and shop at Whole Foods because they need to know that they're helping, and not hurting, the world.

This display of altruism, albeit selfless, has immense benefits to the giver that's been coined as the "helper's high." When we do good, our body releases serotonin, the mood-boosting chemical produced in the gut, which makes us feel happy. This feel-good high can have long-term effects on your body like decreased anxiety and depression, minimized physical pain, strengthened immunity, and improved sleep.

*"When you know yourself, you are empowered.
When you accept yourself, you are invincible."*
TINA LIFFORD

DO THE WORK

I AM...

Together, these are the two most powerful words in the English language. What are you? Write out and complete this sentence twenty-five times with all the things that you are!

SHIFT THE WORK

WORK AT WORK

A PURPOSE-DRIVEN MENTALITY reminds me of
how the army breaks down young recruits, only to slowly
build them back up mentally, physically, and emotionally.
Meanwhile, they're shedding the extra weight in the gut as they
prepare themselves to become part of something greater than
themselves. Our soldiers aren't unfeeling mercenaries. They're
people whose values align with the country whose flag they're
fighting under. Is there a same level of compatibility between
your values and the organization's mission for which you work?
Can you confidently say that you feel like you're part of a greater
purpose?

WILLINGNESS TO SACRIFICE is about not being afraid of how our opinions will make us look. Are you hesitant to take contrarian views if they may separate you from the group and damage relationships? Openness and honesty are about the courage to break free from these restraints of self-interest. Why should we stay silent if we feel the organization is moving away from its mission and values? Everyone says they want to be a part of something bigger than themselves until it comes time to sacrifice. We should be willing to sacrifice our time, our money, and our egos for a job we believe in. The gut brain will suggest the road less traveled, but we must have the courage to take that path even if it means walking alone.

IMPACT-DRIVEN is about whether you feel like what you're doing every day has a meaningful impact. My mother guided me to lead a life where the fire burned bright inside of me. Working for a big organization wasn't going to light me up, which is why I turned down the job at Andersen Consulting and, instead, started SHIFT. It's been our mission since the beginning to show other companies how they can have an impact on the world and their workforce. Meanwhile, we've been our own test case, which is why we take great pride in being named an Inc. Magazine's Best Places to Work for 2017.

STRATEGIES:

YOU. DRIVE. PURPOSE.

Know with your head.

Own with your heart.

Drive with your gut.

This is how you work on your purpose.

"Beginnings are usually scary, and endings are usually sad, but it's everything in between that makes it all worth living."

BOB MARLEY

DO THE WORK

YOUR OBITUARY

This is an effective exercise to help us lead a life without regrets. The name of the exercise says it all.

Grab a sheet of paper and write the obituary you'd want your family, friends, and peers to read the morning after your death. Write down the accomplishments you hope to achieve. Do you want charity and community service to be a major part of your legacy? Do you want to be known as an innovator at work? List the family members you will leave behind. Do you see grandchildren in your future?

Hopefully, when you read over the final version, you'll feel that there's a dimension to life greater than material and personal gain. Start living as this person every day.

Impact painting is similar to the obituary exercise. We did it at my company several years ago. It starts with a short lesson from an art teacher that covers the basics. Then, we set everyone up with an easel and canvas and ask them to paint what an impactful year looks like to them. Someone in our company drew a Buddha. Another person painted a chain of people circling the world. One team member did a drawing of an enormous heart that filled every inch of the canvas with shades of red.

Both these exercises are asking the fundamental question connected to the gut brain: **WHAT IS THIS ALL FOR?**

WORKPLACE

The emergence of social entrepreneurship has been a positive development in recent years. Companies have managed to influence real change in their local communities. Still, it's too little and too dispersed to have an overwhelming impact. The volunteer work and charity is usually enacted as separate from the work of the company. It's time for companies to extend their reach by integrating social responsibility into their work.

In 2014, a special working group of the United Nations drafted a list of seventeen Sustainable Development Goals (SDGs). "The Elders," a group of luminaries that includes Sir Richard Branson, Nelson Mandela, and Jimmy Carter, helped compile this list of the greatest issues facing the planet and humanity. The goals are as follows:

If every organization begins operating with these goals in mind, then we can prevent disaster on a massive scale and find ourselves living on this planet a little longer. Imagine the

SHIFT THE WORK

possibility for change if each company were to select only one item on the list as the focus of their corporate responsibility initiative.

The transformation will begin when the company's impact mission is weaved into its business model. Our mission at SHIFT is to shift the engagement of the workforce from 70 percent disengaged to 70 percent engaged. If we can reach that goal with just our clients, our overall impact on the world will be huge.

Weaving mission into the business model is the difference between one life insurance company simply trying to sell more policies and a different firm that tries to find the right policies for families. It's a bank issuing mortgages they know can actually be repaid. It's about changing the conversation in companies, so the shareholders aren't only asking whether a company is profitable, but also whether it's a force for good in the world.

Organizations should focus on more than just what they do, according to Simon Sinek, in his TED Talk *Start with Why*. They shouldn't be focused on how they do it either. Instead, the only question they should ask is why. As Sinek puts it, Martin Luther King Jr. didn't have a plan. He had a *dream*.

The Purpose Hotel is a Kickstarter-funded project whose goal is to design a hotel where each aspect of the hotel and a guest's stay is designed to benefit others. Internet fees fight human trafficking. Soaps, shampoos, and furniture are purchased from companies with the goal of making a difference in the world. Each one-night stay has the potential to positively impact the lives of a hundred people in the world.

DO THE WORK

PURPOSEFUL PROCUREMENT

List your company's five most consumed products and services, and five purpose-driven vendors you could begin to source from.

PRODUCTS AND SERVICES

1.
2.
3.
4.
5.

VENDORS

1.
2.
3.
4.
5.

SHIFT THE WORK

"The only limit to your impact is your imagination and commitment."

TONY ROBBINS

STRATEGIES:

COMMUNITY

ICE BUCKET CHALLENGE

The Ice Bucket Challenge was pivotal in promoting awareness of the disease amyotrophic lateral sclerosis, or ALS. The idea was simple: film a video of ice being dumped on someone's head in order to promote awareness of the disease. When celebrities began filming themselves taking on the challenge, it became a social media phenomenon. Everybody wanted to participate. Surprisingly, people also wanted to give to the cause. The ALS Association ended up raising $115 million from 2.5 million people. 50 percent of the people who gave were first-time donors. The success of the Ice Bucket Challenge proves that, deep down, everyone likes feeling good about doing good.

THREAD

The social fabric of our communities is built on the family structure. Thread, an organization in Baltimore, engages underperforming high-school students by providing a family of volunteers to help with academic and personal growth. Instead of one person mentoring a student, an entire family is recruited for each child. Someone takes the child to school. Another person picks him or her up. One volunteer is in charge of making sure he or she eats. The program has succeeded in graduating more than 250 students. It's the first program I've seen to raise graduation rates. The founder of the program started with eight hundred medical students at Johns Hopkins. Not a group of people who exactly had a lot of free time. But they made the time.

RHYTHM

As part of our effort to weave impact into our mission, we used to hold what's called "Impact Day," where we got out of the office, twice a year, to spend the entire day servicing the community. Admittedly, we stole the idea from Starbucks, who in 2008 was struggling through the recession. Howard Schultz, who had recently returned as CEO, took ten thousand managers down to New Orleans. Half the day was devoted to meetings, and the other half of the day was spent building houses. Over the years, we've increased our giving efforts both in frequency and in reach. We meet with cause partners throughout the year, on a monthly basis, to learn how we can show up and give back to our local communities. We partnered with an organization called Buy One, Give One (B1G1) to give back to our world and to those who need it most.

When we think about the big issues, the ones that seem almost impossible to tackle, it puts our work into proper perspective. The work issues that seemed unmanageable become painless. It's like looking back at a time in your teens when you were hysterical over breaking up with a boy or girl. At the time, you thought the world was ending. Really, it was just beginning.

On Slack, our company has a channel called #impact. It's a place where members of the team can share ways they are giving back to the community and world. Someone may post about a race they're running, or an upcoming

SHIFT THE WORK

charity event. People post pictures of working at a soup kitchen or spending a week down in New Orleans helping build houses with Habitat for Humanity. It has an amazing ability to encourage people to join their colleagues or find impact activities of their own that better suit them. This goes back to the Ice Bucket Challenge. People want to get involved, but they don't always know the way in. They need a coworker to bring them along to an event or draw attention to an innovative charity that speaks to the person's gut.

SOLUTIONS:

TOOLS, TECHNOLOGY, AND TRAINING

STEPS: YOUR IMPACT

What do you want to be known for at the end of your life? When people evaluate your life, how do you want them to judge your relationships with your partner, children, community, company, coworker, and even random strangers? We all have different answers to these questions. The answers may even change from year to year.

> **BY STAYING ATTUNED TO OUR GUT BRAINS, WE CAN IDENTIFY OUR GUIDING PRINCIPLES AND LEAD LIFE WITH INTENTIONALITY AND DELIBERATENESS.**

If we don't impose our principles on life, then life will inflict its misfortune on everyone.

CHANGE THE WORLD?

An Olympic runner will train ten hours a day. The facility the athlete uses for training is carefully chosen. It needs to have the proper equipment and provide a comfortable environment. The athlete can't be distracted by small inconveniences. Every moment of the day is carefully planned out. The athlete follows a special diet, not consuming any food or drink that will prevent the runner from maintaining peak performance. There is deliberateness to every action. The athlete must maintain this deliberateness even after the training session is over. The Olympic runner doesn't return home and eat a bag of Doritos and drink Mountain Dew. If the athlete doesn't get a good night's rest, then he or she won't be able to train the following morning.

If you want to become a world-class performer, work is your gym! It's where we spend the bulk of our day. When we connect to our bodies at work and hear the signals being sent by the brains in the head, heart, and gut, we can become more innovative, connected, and passionate human beings! When we work on purpose, we go home with purpose too.

Workers don't innovate if they don't feel a passion for the work. Likewise, they don't have a passion for the work if they aren't in a position to innovate.

A DISENGAGED WORKFORCE IS HURTING THE BOTTOM LINE OF COMPANIES AND WORKERS.

In recent years, companies have begun paying closer attention to how values impact their bottom line. They began recognizing the utility of hiring women and minorities. Part of this was because of consumer demand, but people wanted to work in diverse companies. It's not appealing to go to a place that's known for sexism or racism. With these advancements, workers began to see themselves as agents of change. It's empowering for people, and it makes them feel committed to their jobs.

SHIFT

Are you starting to feel ready to listen to the brains in your head, heart, and gut? Ready to become engaged? Great! Now comes the task of shifting your workplace.

It's not a question of can. It's a test of will. We know we should be going to the gym. We know we should be thinking about helping our neighbor. We know we should demand that our bosses adhere to their stated values and mission.

HOW WILL WE TRUST OUR GUTS, FOLLOW OUR HEARTS, AND STICK TO THE PLANS IN OUR HEADS?

CHAPTER 7

SHIFT HAPPENS, WILL YOU?

WHAT NEXT?

The diagnosis is clear. More than two thirds of the workforce isn't engaged, and unless we begin listening to the three brains in our bodies, we will fail at finding the greater purpose to shift these numbers.

In my job as a consultant, I take great pride in making sure my clients walk away with not only fresh insight but also tangible steps they can take to pursue a solution. Success, in other words, isn't achieved when the diagnosis is delivered. If I can't provide the client with effective actions, then my advice isn't worth the paper it's written on. The question for us is how to turn our awareness and knowledge of our biology and its impact on engagement into actionable steps that will shift the direction of our lives. This book has provided tools, tactics, and practices to harness all three of your brains, but a complete shift and lasting commitment will require an overall change in outlook. Implementing the following six-step process will empower you to change your beliefs, behaviors, habits, standards, and results.

SHIFT YOUR PERSPECTIVE

Throughout this book, we've discussed how language has the power to shift perspective. Events are not happening to you but for you. Responsibilities are not something you have to do but get to do. Your greatest hardships and difficulties are nothing less than opportunities to open your eyes to what is important in life!

Parents feel exhausted and stretched at the end of a long day. The last thing they want to do is bathe an irritable child who is running on fumes. The night hinges on the parent's ability to pause and articulate that this isn't a burden but a chance to bond with the child, to make it a loving moment between parent and child and not one that is bitter and full of tears.

*"Life is truly reflection of what we allow
ourselves to see and be."*
TRUDY SYMEONAKIS VESOTSKY

DIG DEEP DOWN

Write down the twenty-five things in your life that truly matter. This is your last will and testament of what is important in life, the goals and ambitions that will bring feelings of regret if not realized by the time you reach the twilight of your days.

If you find yourself struggling with this exercise, consider volunteering in a retirement home. Sit with some of the older residents, and listen to them talk about their lives. You'll hear stories of things left unsaid and unfulfilled dreams. You'll be able to feel their pain over the time that has passed and cannot be brought back. Regret is the greatest hurt because it hits with the realization that something cannot be undone.

The point of the exercise is to direct us to the matters, responsibilities, and people in our lives that we should honor and appreciate. It allows us to identify the superficial noise in our lives that is better left ignored, like the missed parties, the cars we can't afford, and the television shows that we don't have to binge-watch. We're our worst selves when we covet material objects more than time and relationships. In my family, we've adopted a rule of no more gifting objects. The only presents we allow are experiences that allow the family to spend time together, whether it's a vacation, an art project, or dinner out.

Look at your list of the twenty-five things that truly matter, and ask yourself what you have control over and what you don't have control over. Buddhist tradition puts great emphasis on the idea of remaining unattached to the potential outcomes of matters we can't control. We can go to the gym religiously, hire the best financial planner in the world, and buy a lottery ticket every day of our lives, but, ultimately, death, taxes, and winning the lottery are beyond our control. Limited control is a fact of life. We don't have veto power over who will love, marry, and accept us. All we control is the knowledge we let into our minds, the outlook we use to engage the world, and the kindness we show as a parent, spouse, child, sibling, and neighbor.

WE DETERMINE ONLY OUR OWN CHARACTER, HABITS, BEHAVIOR, THOUGHTS, AND WORDS.

If we stay focused on chasing our passions and realizing our dreams—if we keep our eye on the prize—then we'll never feel regret again.

STEP 2:

ESTABLISH A (REAL) PLAN

Everyone sets a New Year's resolution, but few people draft a well-defined plan to help them realize their goals. Instead of ready, aim, fire, it ends up being fire, ready, aim.

Stephen Covey, the author of *The 7 Habits of Highly Effective People*, writes that once the end goal is clear in the person's mind, he or she needs to go through a process of reverse engineering to see whether the goal is realistic, and how it can be broken down into achievable steps.

In high school, I had the thought that I'd play in the NFL one day. I stood at 5'8, 176 pounds. Reverse engineering would've cued me into the reality that I'd have to figure out how to grow six inches, gain a hundred pounds, and shave a second off my forty-yard dash in the next five years leading up to the draft.

If we're going to be **ALL IN**, we need to know what **ALL IN** looks like, and for each person it's different. Most people say they want to lose weight and come up with a random goal of losing a pound a month. These people are playing checkers, thinking three steps ahead, instead of playing chess and breaking down every step needed to clinch victory. It's the difference between high performers and those who whine and cry about not hitting their goals.

I've set, for example, the goal of achieving 7 percent body fat. After talking to people who've achieved a similar goal, I determined, through reverse engineering, the precise discipline and progression that will be required in terms of diet, exercise, and sleep in order to reach my goal.

Having a plan means writing it down. People who write down their goals have a greater chance at succeeding than people who keep them in their heads. It's about sitting down and recording in a journal a realistic, specific, measurable goal that can be executed within a particular time frame.

Establishing a plan may sound basic and simplistic, but it's common sense that's not commonly deployed.

STEP 3:

CELEBRATE YOUR PROGRESS

It's unrealistic to think we can live life on a constant high. Even if we're making progress, there will be a moment when we take a big hit and get knocked down to reality, or our past selves. How we respond to such adversity is a secondary challenge of growth.

In today's culture, we've learned to deal with adversity by taking our ball and going home. We've become a nation of quitters. We've cited the high divorce rates. We've examined the tendency of people to discard a slightly old iPhone in favor of the latest model. Too many of us face resistance, and instead of regrouping and moving ahead, we allow the setback to turn incapacitating.

A person makes a New Year's resolution to stop smoking. She sails through the first three weeks of the year without lighting up. One night, she goes to a party. The smell of a cigarette wafts in from the balcony. Minutes later, she is outside bumming a cigarette off a friend. The next day, she buys a fresh pack.

If only this woman would've stopped and celebrated her progress. Making the resolution is progress in itself. People who explicitly make resolutions are ten times more likely to attain their goals than people who don't explicitly make resolutions. She didn't stop and congratulate herself for sailing through the first three weeks and set a new goal of making it to four. Instead, she turned a small stumble into a complete collapse. She allowed perfection to become the enemy of good.

Scarcity isn't something we suffer from any longer. We lack for nothing. We're past the age of one-size-fits-all answers. Today, the resources to help people overcome their problems are immense. If we want to salvage a relationship, for example, there are a multitude of counselors and therapies to choose from. Instead of chucking our phones at the first sign of wear and tear, we can make use of companies like www.fixt.co that will come to our offices and perform on-site repairs of our broken mobile devices.

Instead of quitting, learn how to fail forward. ***Success is only linear in fantasies.*** The stretch of road that takes us from point A to point B is full of twists, turns, potholes, and bathroom breaks. What's important to remember is that the obstacles only come once we've started the journey. They only happen because we've succeeded in getting past the starting line. They're the greatest proof that progress has been achieved. These unexpected stops are teachable moments when we can modify and sharpen our plans, so we can better navigate the rest of the journey.

Most successful start-ups, in an effort to innovate, develop, and grow their businesses, use a methodology of build, test, and measure. Each step is an integral part of being ALL IN on an idea. Even if the test fails, it's crucial to measure all the ways it has succeeded, so the baby isn't thrown out with the bathwater when version 2.0 is built.

Brothers Dan and Chip Heath write about the need to celebrate progress in their boo, *Switch: How to Change Things When Change Is Hard*. Great leaders, according to the authors, begin all efforts of problem-solving by finding the bright spots of the situation, rather than immediately focusing on the problems. It's about seeing what works so we can better understand how it went wrong.

In therapy, this is called the sandwich technique. The patient begins the session with positive news, and then moves on to something negative, before finally closing with a second positive report.

By nature, we like to present pessimism. Walk down the street and eavesdrop on people's phone calls. An overwhelming majority of the conversations focus on the negative actions of others. Someone is complaining about a spouse, parent, boss, landlord, or politician.

Even if we like spewing pessimism, we much prefer hearing optimism coming out of other people's mouths. Most of us can't stand listening to negativity for more than a couple of minutes. If we stopped and listened to our own thoughts, we wouldn't enjoy it either.

Stand in front of a mirror and speak any negative thoughts aloud to yourself. You'll find that it's not much fun to hear yourself talk about all the ways you've managed to disappoint yourself in the last day.

SHIFT YOUR LANGUAGE, SHIFT YOUR PERSPECTIVE.

Now, stand in front of a mirror and speak any positive thoughts aloud to yourself. Boast to yourself about all the great developments in your life. Celebrate the progress you've made.

STEP 4:

GENERATE PASSION

Fake it to make it.

The way we move and hold our bodies has a direct impact on how we feel about ourselves. If we walk with confidence, we will feel confident. If we slump our shoulders and drag our feet, we'll feel lethargic. It's why it's physically impossible to be mad if you smile, look up at the sky, and jump up and down. Communication, as we've discussed, is mainly about voice quality—tone, pitch, pace—and body language. It's our bodies, not our thoughts or words, that play a central role in generating passion.

PASSION COMES DOWN TO ONE SIMPLE QUESTION: WHY DO WE WANT TO DO THIS?

Once we know the answer, we need to set upon the task of creating the urgency and emotion it will take to make our visions a reality. It begins with the energy we exhibit when we wake up in the morning. It's how we carry our bodies as we walk through our day and the tone of our voice when we speak. Is the face we show the world one that beams, or are our lips turned down in a permanent scowl? When we exercise and get our hearts beating fast, we don't simply feel healthy—we feel optimistic about life. After an hour at the gym, we're more inclined to tackle a life-affirming activity—family time, work, studying—rather than wanting to vegetate in front of the television or go to sleep.

In these moments that we're walking, talking, and speaking with excitement and optimism, we're stuck in the present. We're not dwelling on the past, or anxious about the future. We're not stuck in *a* moment. We're stuck in *the* moment.

In the book *Flow: The Psychology of Optimal Experience*, Mihaly Csikszentmihalyi talks about the power of getting into a state of flow, or what others call the "zone." It's a moment when we have clarity of thought. The world, in effect, slows down.

Baseball players say that when they're in the zone, the ball comes toward the plate in slow motion. It can be a ninety-five-mile-an-hour fastball, but the hitter can see the actual rotations of the ball and its stitching.

Passion is in the eye of the beholder. But if we're failing to clearly convey our passion to others, we're most definitely failing to communicate it to our own three brains.

Tony Robbins says in his talk, *Getting Stuck in the Negatives and How to get Unstuck*: "**MOTION CREATES EMOTION.**"

STEP 5:

CHASE YOUR PURPOSE

Finding your path takes real work. We shouldn't expect to wake up one morning with a sense of purpose. Epiphanies do happen, but only once we've provided the brain in the head with enough thoughts, images, and experiences to make innovative connections. Steve Jobs's unique aesthetic values for Apple products can be traced back to a calligraphy course he took with a monk after dropping out of college. Larry Page came up with the distinctive approach for Google's algorithm after hearing an Italian professor speak at a conference in California.

We must chase our purpose with fervor. When we try to identify what excites us, we shouldn't measure the path simply on how it personally benefits us. This is not the key to finding something sustaining. The litmus test for what's considered a purpose worth going ALL IN for is something that benefits our families, our communities, the entire world, and us. The greatest influencers are people who've managed to come up with a mission that takes all of those elements into account.

What we are trying to find is our mission statement. Why are we here? How do we want to be remembered? If we can't connect our actions and work back to this mission statement, then we shouldn't be doing it.

COMMIT TO PERFORMANCE

Day-to-day life is about routines. We brush our teeth and get dressed a certain way. Every morning, we drive to work along the same route. The way we defuse conflicts fits into certain patterns. Following well-known practices provides us with comfort and certainty, even when they're ineffective. Such rigidity is not a recipe for growth.

Shifting behavior ultimately demands changing patterns. When we act deliberately and intentionally to change our behavior, new neural pathways that connect the three brains are created. These fresh connections better allow us to connect our biology to the mission we've set out to achieve.[CH7:N1]

In order to produce more effective patterns, we first have to recognize our existing ones. **Which patterns are constructive, and which ones are destructive?**

My morning routine, for example, is what I consider constructive. I'm up at 4:30 every morning. The day starts with journaling, followed by writing and drawing. Then, I take a shot of Bulletproof Coffee before going off to boxing class. I'm back home at 6:15 a.m. to shower. Once dressed, I send my daily CEO riff out to the company. Having concentrated on my body, balance, business, and being, I've established my passion and purpose for the day, and I'm now ready to help get the kids ready for their day. This pattern is a keeper. Having recognized this as a constructive pattern, I can set it on rinse and repeat.

My evening routine, on the other hand, is destructive. After having skipped lunch, I return home and gorge. Dinner is incomplete without a bottle of wine and a giant chocolate-chip cookie for dessert. Despite knowing that I'll have to be up before the crack of dawn, I stay up until around 10:30 p.m. It doesn't take a rocket scientist to understand that my nighttime routine needs to go.

How many of us are stuck in a nighttime routine where we fall asleep with the television on and our phones in our hands, even though studies show that nothing could be worse for our sleep patterns? If we stopped to consider how we are harmfully stimulating the brains in our heads, hearts, and guts at a moment when we should be trying to relax our bodies, we'd look for an opportunity to come up with a more positive routine.

We can't abandon our current patterns, unless we have new patterns to take their place. It helps to think of patterns as takeoffs and landings to the crucial activities in our lives.

Say there is a major deadline on a project at work. Normally, you procrastinate and the pressure of the deadline slowly builds. Also, you resist bringing in certain people for advice because you like full control. As the deadline draws closer, your stress level builds. You dread sitting down and doing it. Why not put a process in place to help you start (takeoff) and finish (landing) while maintaining a high level of engagement and a low amount of aggravation? Ask yourself: "What steps have proven effective in getting me to sit down and do the same work in the past? How can I set myself up to constantly succeed?"

THROUGH EFFECTIVE ROUTINES, WE CONTROL THE TONE, MINDSET, AND STANDARDS WE CARRY INTO EVERY MEETING, PROJECT, AND RELATIONSHIP! AT THE VERY LEAST, MAKE THE EFFORT TO DESIGN PURPOSEFUL MORNING AND NIGHTTIME ROUTINES FOR YOURSELF.

COMMIT.

CHAPTER 8

YOUR INNOVATION, INSPIRATION, AND IMPACT

My late father-in-law was a gentle giant, a family man, intensely disciplined. When I think about what it means to go **ALL IN**, it's hard not to think of him. A majority of his adult life was spent carefully planning an early retirement at the age of fifty-seven, even though he wasn't a high earner. At the end of every year, he'd check on his progress, and knock another year off his countdown to retirement.

Shortly after turning fifty-four, the doctors diagnosed him with an impending expiration date. It came right as he was nearing the completion of his life goals. He'd sent his three children to private school and college. The mortgage had been paid off, and he was three years away from closing the work chapter of his life.

There was a time, around three years into his illness, when he was undergoing a particularly heavy course of chemotherapy. One day, when I was getting ready to leave his house after a visit, he stood up and hugged me. He squeezed hard, and it felt like he was using every last ounce of strength in his battered body. Until then, I never thought of him as much of a hugger or kisser, at least with me, his son-in-law.

As he was pulling away from the embrace, he spoke into my ear. "Cancer," he said, "is the best thing that's ever happened to me and certainly the worst. The worst is that I won't live long enough to see my grandchildren grow up. But I've never enjoyed my life more since I've been diagnosed with this terrible disease. It's forced me to change my perspective in a way that I could've never done on my own. It's allowed me to appreciate everything in life on a much higher level."

During the remainder of his illness he hugged everyone a little tighter and often told them how much he loved them. Every funny moment was appreciated a little more. Holidays were a time for the entire family to come together and celebrate. He used the time he had left to strengthen his relationships. Everyone was greeted with a smile. Anger or disapproval never left his lips. He ended up living for another six and a half years.

THIS, TO ME, IS LIVING AN INTENTIONAL LIFE.

You and I don't have to wait for a terminal illness to appreciate the gift of life. The question is: *Are you happy with how you're using yours?* Eventually, we'll all have to face the end of life, so we should constantly be grateful for its great gift. At this moment, the earth and humanity are on the precipice of both major challenges and exciting breakthroughs. The world needs us to act with wider perspective and greater sense of purpose.

Changing weather patterns from climate change are leading to drought and famine around the globe. Forty percent of India doesn't have access to clean drinking water. In many parts of the earth, the shortage of food is causing political instability.

The United States is also facing immense challenges. From 1999 to 2012, the percentage of Americans taking antidepressants grew from 6.8 percent to 13 percent. Ironically, two-thirds of these medicated people are not even clinically depressed.[CH8:N1]

In the United States Americans dispose of more than four hundred pounds of food per person per year. This equates to 1,250 calories per person per day, or enough to feed every man, woman, and child in the country.[CH8:N2]

Despite these many problems, most of us don't feel that we can affect change. Voter turnout for presidential elections in the United States hovers close to 60 percent. For midterm elections, the rate is closer to 40 percent.

This isn't to say we aren't making progress in the fields of science, technology, and medicine.

Elon Musk is working to put humans on Mars in the coming decade, while continuing to grow the electric car market and reduce our dependence on fossil fuels. More and more people are experiencing cancer as a chronic condition instead of a death sentence. In a matter of years, drones will be able to regularly access remote regions of the world, delivering food and healthcare.

However, if 70 percent of us continue to remain disengaged at work, we will fail to build on these advances or channel them to positive ends. If we don't manage to shift the workforce, we'll get caught behind the eight ball. The regrets we'll have will not be just as individuals but as a country. We have the greatest potential to continue to wave the flag of freedom and solve the scourges of poverty, inequality, and racism. This opportunity will slip through our fingers if we continue wasting time on Facebook, binge-watching shows on Netflix, or buying gifts for people who don't need them. Most crucially, we will fail to reach our potential if we stay in jobs we don't love because of money or a need to impress others.

THE WORLD NEEDS YOUR INNOVATION, INSPIRATION, AND IMPACT, AND IT NEEDS IT NOW.

The decision to go **ALL IN** is neither dogma nor motivational claptrap. The science tells us we can integrate our head, heart, and gut into the work world by thinking in a deliberate way, using certain words, and creating habits and patterns of behavior. If we can use the three brains to become fully engaged for the eight hours a day we spend at work, if we can generate passion and purpose and establish constructive patterns, then we'll return home as better spouses, parents, children, and citizens of our communities and the planet. This is how we'll get rid of the current apathy that has us playing the part of backseat drivers while the world suffers. Deep down we all have the potential to become superheroes, but it starts with a deep alignment of meaning.

We're hearing whispers of radical and rapid change because there's profit to yield from treating workers and their values with respect. We can build on this progress if more people learn to respect themselves and believe in their ability to impact the world. Protesting is effective, up to a point. Policy, however, is changed only when people make different choices.

"I AM WHO I AM BECAUSE YOU ARE WHO YOU ARE."

IF WE PERSONALLY COMMIT TO BEING ALL IN, THEN WE'LL HELP THOSE AROUND US ARRIVE THERE TOO. LIKEWISE, IF WE FAIL AT IT OURSELVES, THEN IT WOULD BE FOOLISH AND HYPOCRITICAL TO EXPECT IT OF OTHERS.

CHAPTER 9

THE DARE

WE SHIFT THE WORK, THEN THE WORK SHIFTS US

My life is full of experiences that I never thought were possible or available to me. Experiences that other people took for granted always seemed beyond my reach from where I was standing, which was generally on the outside looking in. Everything happened because I was willing to take that first step on the path of the road less traveled.

One of my earliest memories of standing outside looking in happened in high school. Walking through the school one day, I heard music coming from down the hallway. Through a classroom door window, I could see the school band practicing, led by a tall, thin African American called Mr. Harry Watkins. Music had always been a passion. I'd been playing percussion and drums my entire life. I was drawn to the sounds emanating from the room, and every day I'd spend the period walking past the room, envious of my schoolmates. At that moment, I realized what a downer it was not having this musical outlet in my life. The problem was that I played football, and at the roughest high school in Baltimore City, football players weren't "band geeks." It was that simple.

One day, the band director noticed me standing at the window. He wanted to know my deal, why I spent my time stalking the band's rehearsals. We had a brief conversation at the door. I told him I played, but had no interest in joining the band. Football was the excuse I gave him, and he immediately understood my predicament.

Mr. Harry Watkins dared me to join the band, knowing that, as a jock, I'd have a tough time turning down a dare. His goading worked, and the next time I came to that door, I walked right through it and took a seat with the other players in the band.

Mr. Harry Watkins dared me to lead the percussion section. I led.

Mr. Harry Watkins dared me to learn the set on my own. I did.

Mr. Harry Watkins dared me to join the national honor band even though only six percussionists from across the country would be selected. He helped me submit my tape, and I made the cut. Months later, along with the most skilled junior musicians in the country, we put on the greatest concert the University of North Carolina at Greensboro has ever witnessed.

Mr. Harry Watkins, in a challenge that would have the most profound impact on my life, dared me to apply to Johns Hopkins University. I applied. Months later, the acceptance letter came in the mail, and the rest, as they say, is history.

Mr. Harry Watkins was one of the greatest teachers in my life. He knew how to light a fire under me. He put me in touch with my inner superhero.

Unfortunately, the story doesn't have a happy ending. During my senior year of high school, Mr. Harry Watkins grew gravely ill. In his final weeks, he invited me to visit him, but preoccupied with my mother's illness, I kept pushing off plans to see him. As these stories go, when I finally came to terms with my foolishness and made the effort, I was an hour too late.

To this day, I continue to hear his message that every day counts, and that I live life to the fullest by daring myself to push ahead and go ALL IN. To stop being afraid to reach my potential. Now it's my turn to share this gift with you.

I DARE YOU TO LOSE YOUR EGO.

I DARE YOU TO NOT FEEL ENTITLED.

I DARE YOU TO LISTEN TO YOUR HEAD, HEART, AND GUT.

I DARE YOU TO DO WHAT LIGHTS YOU UP.

I DARE YOU TO SHIFT YOUR WORK.

I DARE YOU TO GO ALL IN!

THE ALL-IN
PLEDGE

We believe that all of us can and must...

1. GROW REGARDLESS!

No matter the circumstance, we find a way to take ownership of moving forward. It's on us to be responsible and accountable. Our scars and struggles build character and a foundation to live our greatest purpose in life.

2. GO ALL IN!

It's time to unleash the power of our head, heart, and gut on business to engage more fully in every moment of life. Our connectedness will influence our purpose and drive, emotion and passion, and perspective and priority, leading to enhanced focus and impact.

3. DO THE WORK!

We are the product of the eight (or more) hours we spend at work every day. Only we can make them count. Our craft—what we do, how we do it, with whom we do it—is a reflection of our life and affects our body, balance, being, and brains. Change the work, change our lives.

4. BE BRAVE!

Reveal our truth. To bloom, we need to break open to our core and discover the messy, ugly, and beautiful truths along the journey of life…always and often.

5. MAKE BOLD SHIFTS!

It's time to free ourselves from old thinking, paradigms, and ideologies. We lead by example and empower others to make bold shifts. In our fearless execution, we kick ass, take names, and leave others wondering how we have such massive impact.

Take the pledge at shiftthework.com/all-in-pledge.

CONCLUSION

THE INVITATION

BY ORIAH

It doesn't interest me what you do for a living. I want to know what you ache for and if you dare to dream of meeting your heart's longing.

It doesn't interest me how old you are. I want to know if you will risk looking like a fool for love for your dream for the adventure of being alive.

It doesn't interest me what planets are squaring your moon… I want to know if you have touched the centre of your own sorrow, if you have been opened by life's betrayals or have become shrivelled and closed from fear of further pain.

I want to know if you can sit with pain mine or your own without moving to hide it or fade it or fix it.

I want to know if you can be with joy mine or your own if you can dance with wildness and let the ecstasy fill you to the tips of your fingers and toes without cautioning us to be careful to be realistic to remember the limitations of being human.

It doesn't interest me if the story you are telling me is true. I want to know if you can disappoint another to be true to yourself. If you can bear the accusation of betrayal and not betray your own soul. If you can be faithless and therefore trustworthy.

I want to know if you can see Beauty even when it is not pretty every day. And if you can source your own life from its presence.

I want to know if you can live with failure yours and mine and still stand at the edge of the lake and shout to the silver of the full moon, "Yes."

It doesn't interest me to know where you live or how much money you have. I want to know if you can get up after the night of grief and despair weary and bruised to the bone and do what needs to be done to feed the children.

It doesn't interest me who you know or how you came to be here. I want to know if you will stand in the centre of the fire with me and not shrink back.

It doesn't interest me where or what or with whom you have studied. I want to know what sustains you from the inside when all else falls away.

I want to know if you can be alone with yourself and if you truly like the company you keep in the empty moments.

ACKNOWLEDGMENTS

The acknowledgment section of the book is my favorite part because it provides me the chance to appreciate and recognize those who've helped me learn and live to shift the work. It's also my least favorite section because I will either forget someone or fail to properly convey someone's contribution, and unless the reader knows the people included, it reads like one long inside joke. In other words, please read on if you think your name will appear in the coming paragraphs, and, if not, this would be a great place to stop.

Thank you to my family. First, my core family. If the nucleus is the center of the cell, then this group of people is the center of me and this work. My wife, Erica, is my ground, has my back, and is certainly the love of my life. To my kids, Eliana and James, who are my heart, mirror, and teachers. To my dad, who is my rock and inspiration. To Rose, who is in fact my rose and the most unconditionally supportive person I've ever met. Thank you to my brother, Eddie, and my sister-in-law Kim; my sister, Marisa, and my brother-in-law Trey; my sisters-in-law Megan and Katie; and my mother-in-law, Mickey, for exhibiting boundless love and patience through the years.

Thank you to my work family: Misti Aaronson, Andrew Freedman, and Chris Steer. This section doesn't provide ample space for me to say everything I want about you three. The loyalty you all exhibit toward SHIFT, the love you have for our mission, and the life you've helped us build here at work is nothing I would've imagined when starting this company. I am who I am, because of who we all are. If a person is the average of the five people he or she surrounds him or herself with most, then I'm in good standing with you three at my side. To the rest of the squad at SHIFT: your energy, effort, and enthusiasm for our mission is undeniable and humbling.

Thank you to my mentors who have continued to help me find my greatest strengths, passions, and purpose, and for finding time to invest in me: Harry Watkins, Stan Burns, Steve Lishansky, Coach Roger Wrenn, Chris Gaito, Yanik Silver, Derek Coburn, and many others.

Thank you to the icons I admire most: Richard Saul Wurman for your understanding, Tony Robbins for your relentless passion, Brendon Burchard for your unique ability to unpack the process, Elon Musk for your boldness, Sir Richard Branson for your play, Tupac for your words, Nelson Mandela for your courage and forgiveness, and Sheryl Sandberg for your vulnerability.

Thank you to the folks who make me sound clearer than I talk, and for helping this manuscript evolve into a powerful manifesto on engagement in the workplace: Avner Landes, Tucker Max, Holly Forman, Kate Rallis, and Tara Fox.

A shout-out to my many struggles, bumps, and bruises encountered along the way. Without the many failures and hard lessons learned—sometimes hitting rock bottom—I wouldn't have had the ability to Grow Regardless, Shift the Work, and live out my dharma. Thank you, thank you, thank you.

Don't worry if you ended up deciding to read this section and didn't find your name in it. The show is not over. The movement is just beginning. Like the YouTube video of the guy dancing alone on the street of Liverpool to a busker playing guitar: Soon a young man joins him. Moments later, several more men follow. Then, a group of women start dancing as well. Within a minute, the entire street has erupted into a giant dance party. Forget the acknowledgments.

THE REAL QUESTION IS HOW WE CAN ALL PLAY A PART IN THIS MOVEMENT OF SHIFTING THE ENGAGEMENT OF THE AMERICAN WORKFORCE, SO WE CAN SHIFT THE WORLD.

ARE YOU IN?
ARE YOU ALL IN?

NOTES

INTRODUCTION: THE ENGAGEMENT CRISIS

1. For further research, see the publication, "More than job satisfaction"
    (Vol 44, No. 11) [Internet] by Kirsten Weir (2013). *American Psychological
    Association*, APA.org. Accessible from: http://www.apa.org/
    monitor/2013/12/job-satisfaction.aspx. Data has been sourced from
    "State of the American Workplace" [Internet], report by Gallup Inc. (2013).
    Gallup.com. Accessible from: http://news.gallup.com/reports/199961/7.
    aspx?utm_source=SOAWlaunch&utm_campaign=StateofAmericanWorkp
    lace-Launch&utm_medium=email&utm_content=nonpeek, third iteration,
    2010-2016.

2. See Sarah Gray's report on parts of Baltimore's impoverished economy,
    "6 Shocking Facts About Poverty in Baltimore" [Internet], 28 April 2015.
    ATTN:, attn.com. Accessible from: https://www.attn.com/stories/1541/
    baltimore-poverty-facts.

3. See Amy Adkins, "Millennials: The Job-Hopping Generation" [Internet],
    12 May 2016. Business Journal. *Gallup Inc*. Accessible from: http://news.
    gallup.com/businessjournal/191459/millennials-job-hopping-generation.
    aspx

4. The United States Department of Labor, Bureau of Labor Statistics,
    provides the Economic News Release (2016), "Distribution of employed
    wage and salary workers by tenure with current employer, age, sex, race,
    and Hispanic or Latino ethnicity, January 2016" [Internet] (Table 3) with
    percentage distribution by tenure with current employer, https://www.
    bls.gov/news.release/tenure.t03.htm. *Bureau of Labor Statistics*, Bls.gov.
    Employee Tenure research accessible here: Table of Contents [Internet],
    https://www.bls.gov/news.release/tenure.toc.htm (2006-2016).

5. Median household incomes have fallen significantly for young, educated
    graduates entering the workforce compared to their baby boomer parents,
    this analysis finds: "Millennials earn 20% less than Boomers did at same
    stage of life" [Internet] published by USA Today (2017). *The Associated
    Press*, USAToday.com. Accessible from: https://www.usatoday.com/story/
    money/2017/01/13/millennials-falling-behind-boomer-parents/96530338/.
    The analysis credits this pattern to present diminishing opportunity.

CHAPTER 1: WORK ON PURPOSE

1. See Christian Jarrett, "How Expressing Gratitude Might Change Your Brain" [Internet] January 7, 2016. *New York Media*, NYMag.com. Accessible from: http://nymag.com/scienceofus/2016/01/how-expressing-gratitude-change-your-brain.html for research on positive psychology and the neurological effects of practicing gratitude.

2. Robert M. Sapolsky, *Why Zebras Don't Get Ulcers: An Updated Guide To Stress, Stress Related Diseases, and Coping* [Print]. 2nd Rev Ed, April 15, 1998. W. H. Freeman.

3. Andrew Chamberlain, "What Matters More to Your Workforce Than Money," [Internet] January 17, 2017. *Harvard Business Review*, HBR.org. Accessible from: https://hbr.org/2017/01/what-matters-more-to-your-workforce-than-money.

4. Patrick Wong, "Does More Money Change What We Value At Work?" [Internet] January 17, 2017. *Glassdoor*, Glassdoor.com. Accessible from: https://www.glassdoor.com/research/more-money-change-value-at-work/.

CHAPTER 2: EVOLUTIONARY ENGAGEMENT

1. The United States Department of Labor, Bureau of Labor Statistics released their findings in "Job Openings and Labor Turnover Summary" [Internet] on 7 November 2017, stating 6.1 million job openings were at hand on the last business day of September 2017. Accessible here: https://www.bls.gov/news.release/jolts.nr0.htm https://www.bls.gov/news.release/jolts.nr0.htm. Statistics stating unemployed workers at 6.5 million can be further found here: "The Employment Situation–October 2017" [Internet], accessible from: https://www.bls.gov/news.release/pdf/empsit.pdf.

2. Kathleen McAuliffe, "If Modern Humans Are So Smart, Why Are Our Brains Shrinking?" [Internet] from September 2010 Issue, published online January 20, 2011. *Discover Magazine*, DiscoverMagazine.com. Accessible from: http://discovermagazine.com/2010/sep/25-modern-humans-smart-why-brain-shrinking. Research conducted in conjunction with John Hawks, David Geary, Christopher Stringer, Drew Bailey, Richard Wrangham, Brian Hare, and Richard Jantz.

3. For further results on efforts to reduce poverty rates, see the World Bank's article, "Understanding Poverty," [Internet], last updated 2 October 2016. Accessible from: http://www.worldbank.org/en/topic/poverty/overview.

4. See The American Heart Association's article on "Understanding the American Obesity Epidemic" [Internet], 9 March 2016. The article directly states that "Currently, one in three U.S. adults is obese." Accessible from: http://www.heart.org/HEARTORG/HealthyLiving/WeightManagement/Obesity/Understanding-the-American-Obesity-Epidemic_UCM_461650_Article.jsp#.WhWgobQ-fOR. Further data regarding ranking of health concerns can be found at Gallup's article, "Americans Say Health Is Top Health Problem" [Internet] (2013) which states in the fourth paragraph, "Among actual health ailments, obesity ranks first, with 13% naming it... Simultaneously, cancer, the second-leading cause of death in America, has fallen to fourth place on the list. Two percent mention heart disease, the nation's leading cause of death." While cost and access to healthcare are deemed most detrimental by Americans surveyed, the 'actual health ailment' that Americans have put at the top of the list is the country's rapidly rising rates of obesity. Accessible from: http://news.gallup.com/poll/165965/americans-say-cost-top-health-problem.aspx. You can learn more about the obesity epidemic here: https://www.webmd.com/diet/obesity/features/obesity-epidemic-astronomical#1.

5. See Kimberly Amadeo's "The 5 Most Important U.S. GDP Statistics and How to Use Them" [Internet], 27 October 2017. The Balance. Accessible from https://www.thebalance.com/u-s-gdp-5-latest-statistics-and-how-to-use-them-3306041. For a thorough understanding of World Bank Group's debt-to-GDP ratio calculations and concerns, read "Finding the tipping point -- When Sovereign Debt Turns Bad" [Internet]: https://elibrary.worldbank.org/doi/abs/10.1596/1813-9450-5391. By Thomas Grennes, Mehmet Caner, and Fritzi Koehler-Geib (2010).

6. Bill Quigley, "40 Reasons Why Our Jails Are Full of Black and Poor People" [Internet], 2 June 2016. *HuffPost News*, HuffingtonPost.com. Accessible from: https://www.huffingtonpost.com/bill-quigley/40-reasons-why-our-jails-are-full-of-black-and-poor-people_b_7492902.html. Quigley explores the effects of race and economic status on America's incarceration rates, which have increased from 500,000 to 2.2 million from 1980 to 2015.

7. Peter Wagner, Bernadette Rabuy, "Mass Incarceration: The Whole Pie

2017" [Internet], March 14, 2017. *Prison Policy Initiative*, PrisonPolicy.org. Accessible from: https://www.prisonpolicy.org/reports/pie2017.html.

8. Jacob Masters, "Texting While Driving Vs. Drunk Driving: Which Is More Dangerous?" [Internet], October 27, 2013. *Brain Injury Society*, BISociety. org. Accessible from: http://www.bisociety.org/texting-while-driving-vs-drunk-driving-which-is-more-dangerous/. Data sourced from The United States Department of Transportation, National Highway Traffic Safety Administration (NHTSA), "Distracted Driving". NHTSA.gov. Accessible from: https://www.nhtsa.gov/risky-driving/distracted-driving.

9. Anne Holub, "Why Most People Quit Their Jobs" [Internet] (2015). *PayScale Inc.*, PayScale.com. Accessible from: https://www.payscale.com/career-news/2015/09/why-most-people-quit-their-jobs

10. Neuroscience News summarizes findings from the Salk Institute on dopamine levels' effects on decision-making, claiming there is an impact from our mental pathways to physical performance with trials done on mice. "Hard Choices? Ask Your Brain's Dopamine" [Internet], 9 March 2017. *Neuroscience News*, NeuroscienceNews.com. Accessible from: http://neurosciencenews.com/dopamine-decision-making-6222/. See direct report at "Dynamic Nigrostriatal Dopamine Biases Action Selection" by Christopher D. Howard, Hao Li, Claire E. Geddes, and Xin Jin in Neuron. Published online 9 March 2017. Accessible from: https://www.ncbi.nlm.nih. gov/pubmed/28285820.

11. Zainal Ariffin Ahmad, "Brain in Business: The Economics of Neuroscience" [Internet], 2010 April-June; 17(2): 1-3, The Malaysian Journal of Medical Sciences, United States National Library of Medicine. National Institutes of Health, Ncbi.nlm.gov. Accessible from: https://www.ncbi.nlm.nih.gov/pmc/articles/PMC3216154/

CHAPTER 3: THE FIRE INSIDE

1. John Roach, "Why Fire Walking Doesn't Burn: Science or Spirituality?" [Internet], 1 September 2005. *National Geographic News*, News.NationalGeographic.com. Accessible from: https://news.nationalgeographic.com/news/2005/09/0901_050901_firewalking.html.

2. See Donald R Hoover, Stephen Crystal, Rizie Kumar, Usha Sambamoorthi, and Joel C Cantor, "Medical Expenditures during the Last Year of Life: Findings from the 1992-1996 Medicare Current Beneficiary Survey"

[Internet] Health Services Research v.37(6); 2002 Dec 37(6): 1625-1642. Draws from the 1992-1996 Medicare Current Beneficiary Survey (MCBS) comparing costs of medical expenditures for the elderly (aged 65 and above). Accessible from: https://www.ncbi.nlm.nih.gov/pmc/articles/PMC1464043/

3. Paul Hodgson, "Top CEOs make more than 300 times the average worker" [Internet], 22 June 2015. *FORTUNE*, Fortune.com. Accessible from: http://fortune.com/2015/06/22/ceo-vs-worker-pay/.

4. Valentina Zarya, "These Charts Show Exactly How Few Minority Women Are in Positions of Power" [Internet], 30 March 2016. FORTUNE, Fortune.com. Accessible from http://fortune.com/2016/03/30/charts-minority-women-leaders/.

5. See Rachel Hallett and Rosamond Hutt, "10 jobs that didn't exist 10 years ago" [Internet], 7 June 2016 for the full list of newly-coined positions and their descriptions. World Economic Forum, WEForum.org. Accessible from: https://www.weforum.org/agenda/2016/06/10-jobs-that-didn-t-exist-10-years-ago/.

6. Lydia Dishman, "Scientific Proof That Your Gut Is Best At Making Decisions" [Internet], 31 July 2015. *Fast Company & Inc*, FastCompany.com. Accessible from https://www.fastcompany.com/3049248/scientific-proof-that-your-gut-is-best-at-making-decisions.

7. Amy Fraher, "Creating an emotionally intelligent warrior" [Internet], 9 November 2011. *The Washington Post*, WashingtonPost.com. Accessible from https://www.washingtonpost.com/national/on-leadership/creating-an-emotionally-intelligent-warrior/2011/11/09/gIQA5EOk6M_story.html?utm_term=.7caaca7aa4a4.

8. Katie M. Palmer, "A First Big Step Toward Mapping the Human Brain," [Internet], May 14, 2015. *WIRED*, Wired.com. Accessible from https://www.wired.com/2015/05/first-big-step-toward-mapping-human-brain.

9. See Daniel Goleman, "Emotional Intelligence" [Internet] for further reading on his inspiration from John Mayer and Peter Salovey's first formulation of the concept of emotional intelligence (EQ) as a success factor to accompany the widely-known IQ. Daniel Goleman, DanielGoleman.info. Accessible from http://www.danielgoleman.info/topics/emotional-intelligence/.

10. See "Hugging: A Spiritual Perspective" from the Spiritual Science Research Foundation. Spiritual Science Research Foundation Inc., SpiritualResearchFoundation.org. Accessible from https://www. spiritualresearchfoundation.org/spiritual-living/how-should-we-greet/ people-hugging/.

11. See "Thinking from the Heart - Heart Brain Science" [Internet] from Noetic Systems International on the 'intelligence' of the heart. NoeticSi.com. Accessible from http://noeticsi.com/thinking-from-the-heart-heart-brain-science/.

12. Liz Mineo, "Good genes are nice, but joy is better" [Internet], 11 April 2017. *Harvard Gazette*, News.Harvard.Edu/Gazette. This series by Harvard researchers tackles the issue of aging and provides findings on the correlation between healthy relationships and happy lives. Accessible from https://news.harvard.edu/gazette/story/2017/04/over-nearly-80-years-harvard-study-has-been-showing-how-to-live-a-healthy-and-happy-life/.

13. Carolyn Gregoire, "The 75-Year Study That Found The Secrets To A Fulfilling Life" [Internet] 11 August 2013. *HuffPost News*, HuffingtonPost. com. Accessible from http://www.huffingtonpost.com/2013/08/11/how-this-harvard-psycholo_n_3727229.html.

14. Richard Feloni, "A Zappos employee had the company's longest customer-service call at 10 hours, 43 minutes" [Internet] 26 July 2016. *Business Insider*, BusinessInsider.com. Accessible from http://www.businessinsider. com/zappos-employee-sets-record-for-longest-customer-service-call-2016-7.

15. Jim Edwards, "Check Out The Insane Lengths Zappos Customer Service Reps Will Go To" [Internet], 9 January 2012. *Business Insider*, BusinessInsider.com. Accessible from http://www.businessinsider.com/ zappos-customer-service-crm-2012-1.

16. Brown, Brene. "The Power of Vulnerability" [Internet video]. TEDxHouston, TED Talks, June 2010. Accessible from The Power of Vulnerability. TED. com. Accessible from https://www.ted.com/talks/brene_brown_on_ vulnerability.

17. See Jeff Fromm and Barkley's website MillennialMarketing.com for further statistics on the marketing trends of Millennial consumers (defined as

those born between the years of 1977-2000.) *Powered by FutureCast*, MillennialMarketing.com. Accessible from http://www.millennialmarketing. com/who-are-millennials/.

18. Duckworth, Angela Lee. "Grit: The power of passion and perseverance" [Internet video]. TED Talks Education, April 2013. Accessible from Grit: The power of passion and perseverance. TED.com. Accessible from https:// www.ted.com/talks/angela_lee_duckworth_grit_the_power_of_passion_ and_perseverance/up-next.

19. Burchard, Brendon. "My Car Accident (or, The Power of Grace and Intention" [Internet video]. Brendon Burchard YouTube Channel, 2017. Accessible from Brendon Burchard YouTube Channel. Accessible from https://www.youtube.com/user/BrendonBurchard/ search?query=how+mortality.

CHAPTER 4: TIMELESSNESS WILL SET YOU FREE

1. See Steven Piersanti, "The 10 Awful Truths about Book Publishing" [Internet], 26 September 2016, for even more depressing statistics about the realities of book publishing. *Berrett-Koehler Publishers*, BKConnection. com. Accessible from https://www.bkconnection.com/the-10-awful-truths- about-book-publishing?redirected=true.

2. Laura Helmuth, "Top Ten Myths About the Brain," 19 May 2011. *Smithsonian Magazine*, SMITHSONIAN.COM. Accessible from http:// www.smithsonianmag.com/science-nature/top-ten-myths-about-the- brain-178357288/

3. Gary Vaynerchuk, "You Won the Lotto" [Internet video], 28 September 2016. *YouTube*, YouTube.com. Accessible from https://www.youtube.com/ watch?v=4hHpQP69asA. One of Gary's many motivational videos on his YouTube channel.

4. Ali Binazir, "What are the chances of your coming into being?" [Internet], 15 June 2011. *Harvard Blog*, Blogs.Harvard.Edu. Accessible from http://blogs. harvard.edu/abinazir/2011/06/15/what-are-chances-you-would-be-born/

5. Larry Kim, "Multitasking Is Killing Your Brain" [Internet], 2 February 2016. *OBSERVER*, Observer.com. Accessible from http://observer.com/2016/02/ multitasking-is-killing-your-brain/.

CHAPTER 5: DO YOU THUMP?

1. HeartMath Institute, "The Heart-Brain Connection" [Internet]. *The HeartMath Institute*, HeartMath.org. Accessible from https://www.heartmath.org/programs/emwave-self-regulation-technology-theoretical-basis.

2. Tony Robbins, "Tony Robbins: 6 Basic Needs That Make Us Tick" [Internet], 4 December 2014. *Entrepreneur Media Inc.*, Entrepreneur.com. Accessible from https://www.entrepreneur.com/article/240441.

3. HeartMath Institute, "The Energetic Heart Is Unfolding" [Internet], 22 July 2010. *The Heartmath Institute*, Heartmath.org. Accessible from: https://www.heartmath.org/articles-of-the-heart/science-of-the-heart/the-energetic-heart-is-unfolding.

4. American Heart Association, "Is Broken Heart Syndrome Real?" [Internet], Updated 12 December 2017. *American Heart Association*, www.heart.org. Accessible from: http://www.heart.org/HEARTORG/Conditions/More/Cardiomyopathy/Is-Broken-Heart-Syndrome-Real_UCM_448547_Article.jsp#.WmNM9ainE2w.

5. See Juliette Kando, "Second Brain Found in Heart Neurons - Trust Your Gut Feelings" [Internet], for examples of heart transplant patients who have changed their behavioural patterns and interests, unknowingly adapting to the donors' personalities. They are those who seemingly operate using this 'second brain' post-surgery. 25 March 2017. *HubPages*, Hubpages.com. Accessible from http://hubpages.com/education/your-second-brain-is-in-your-heart.

6. (4) Duff McDonald, "The CEO Factory: Ex-McKinsey Consultants Get Hired to Run the Biggest Companies" [Internet], 10 September 2013. Accessible from http://observer.com/2013/09/the-ceo-factory-ex-mckinsey-consultants-get-hired-to-run-the-biggest-companies/.

CHAPTER 6: NO GUTS, NO GLORY!

1. Adam Hadhazy, "Think Twice: How the Gut's "Second Brain" Influences Mood and Well-Being" [Internet], 12 February 2010. *Scientific American*, Scientificamerican.com. Accessible from https://www.scientificamerican.com/article/gut-second-brain.

2. Helen Fields, "The Gut: Where Bacteria and Immune System Meet" [Internet], November 2015. *John Hopkins Medicine*, Hopkinsmedicine. org. Accessible from https://www.hopkinsmedicine.org/research/ advancements-in-research/fundamentals/in-depth/the-gut-where-bacteria-and-immune-system-meet.

CHAPTER 7: SHIFT HAPPENS, WILL YOU?

1. Vivian Giang, "What It Takes To Change Your Brain's Patterns After 25" [Internet], 28 April 2015. *Fast Company Inc.*, FastCompany.com. Accessible from https://www.fastcompany.com/3045424/what-it-takes-to-change-your-brains-patterns-after-age-25.

CHAPTER 8: YOUR INNOVATION, INSPIRATION, AND IMPACT

1. Justin Karter, "Percentage of Americans on Antidepressants Nearly Doubles" [Internet], 6 November 2015. *Mad in America Foundation*, Madinamerica.com. Accessible from https://www.madinamerica. com/2015/11/percentage-of-americans-on-antidepressants-nearly-doubles/.

2. See Dana Gunders, JoAnne Berkenkamp, Darby Hoover, and Andrea Spacht, "WASTED: Second Edition of NRDC's Landmark Food Waste Report" [Internet], 17 August 2017. *Natural Resources Defense Council*, NRDC.org. Accessible from https://www.nrdc.org/experts/andrea-spacht/report-wasted.

ABOUT THE AUTHOR

Joe Mechlinski, a *New York Times* best-selling author, engaging speaker, CEO of SHIFT, and avid community activist, has a deep-rooted passion for helping to build company cultures based on mission, strategy, and growth, regardless of economic or industrial circumstance. He believes in creating cultures where the best talent will seek out opportunities with the best places to work, because when you create something that attracts talent, the talent will stay.

His team at SHIFT embodies the all-in attitude Joe lives by. SHIFT is a collective of four distinct and powerful businesses centered around a mission to shift the work world to transform the real world. They include SHIFT Consulting, SHIFT Recruiting, SHIFT Society, and SHIFT Ventures.

For SHIFT and Joe, the realization of this mission means creating a more engaged workforce.

One of his biggest passions remains giving back—especially in Baltimore, where he got his start.

Above all, Joe is a devoted husband and loving father to two beautiful children, in whom he is instilling the values of going all in and growing regardless.